DISASTER NATIONALISM

EXPERIMENTAL FUTURES: TECHNOLOGICAL LIVES, SCIENTIFIC ARTS, ANTHROPOLOGICAL VOICES
A series edited by Michael M. J. Fischer and Joseph Dumit

Vivian Y. Choi
DISASTER NATIONALISM

TSUNAMI AND CIVIL WAR IN SRI LANKA

DUKE UNIVERSITY PRESS
Durham and London 2025

© 2025 Duke University Press
All rights reserved
Printed in the United States of America on acid-free paper ∞
Project Editor: Liz Smith
Designed by Dave Rainey
Typeset in Adobe Jenson Pro and Real Head Pro by
Westchester Publishing Services

Library of Congress Cataloging-in-Publication Data
Names: Choi, Vivian Y., [date] author.
Title: Disaster nationalism : tsunami and civil war in Sri Lanka / Vivian Y. Choi.
Other titles: Experimental futures.
Description: Durham : Duke University Press, 2025. | Series: Experimental futures | Includes bibliographical references and index.
Identifiers: LCCN 2024041527 (print)
LCCN 2024041528 (ebook)
ISBN 9781478031635 (paperback)
ISBN 9781478028468 (hardcover)
ISBN 9781478060673 (ebook)
Subjects: LCSH: Indian Ocean Tsunami, 2004. | Hazard mitigation—Sri Lanka. | Terrorism—Sri Lanka—Prevention. | Technology and state—Sri Lanka. | Sri Lanka—History—Civil War, 1983–2009.
Classification: LCC HV603 2004.S72 C46 2025 (print) | LCC HV603 2004.S72 (ebook) | DDC 363.34/94091824—dc23/eng/20250211
LC record available at https://lccn.loc.gov/2024041527
LC ebook record available at https://lccn.loc.gov/2024041528

사랑하는 엄마, 아빠 모든 것에 감사드립니다

Contents

Note on Transliteration	ix
Abbreviations	xi
Preface: The Struggle Endures	xiii
Acknowledgments	xvii
INTERLUDE	1
Introduction	3
INTERLUDE	35
1 **Emergence**	37
INTERLUDE	57
2 **Anticipation**	59
INTERLUDE	77
3 **Endurance**	79
INTERLUDE	101
4 **Reiteration**	103
INTERLUDE	135
Postscript	137
Notes	147
References	159
Index	191

Note on Transliteration

Following conventions of other ethnographies of Sri Lanka, rather than use a standard system for transliteration, I have written Tamil and Sinhala words phonetically in English. However, the English usages I follow are those commonly used for those words (e.g., *prashanai*, the Tamil word for problems). Place names and proper nouns are rendered in the text in the form they commonly take in the Sri Lankan press.

Abbreviations

AAAS	American Association for the Advancement of Science
AMDP	Accelerated Mahaweli Development Project
BBC	British Broadcasting Corporation
BBS	Bodu Bala Sena (Army of Buddhist Power or Buddhist Power Force)
CID	Criminal Investigation Department
DMC	Disaster Management Centre
DRM	disaster risk management
DRR	disaster risk reduction
GIS	geographic information systems
GOSL	Government of Sri Lanka
IDP	Internally Displaced Person
IPKF	Indian Peacekeeping Force
JVP	Janatha Vimukthi Peramuna (People's Liberation Front)
LTTE	Liberation Tigers of Tamil Eelam
NGO	nongovernmental organization
PTA	Prevention of Terrorism Act
P-TOMS	Post-Tsunami Operation Management Structure
RADA	Reconstruction and Development Agency
SMS	Short Message Service
STF	Special Task Force
TAFREN	Task Force to Rebuild the Nation
TULF	Tamil United Liberation Front
UNDP	United Nations Development Program
UNHIC	United Nations Humanitarian Information Centre
UNISDR	United Nations International Strategy for Disaster Reduction
UNOCHA	United Nations Office for the Coordination of Humanitarian Affairs
UTHR-J	University Teachers for Human Rights (Jaffna)
VOC	Vereenigde Nederlandsche Geoctroyeerde Oostindische Compagnie (Dutch East India Company)

Preface
The Struggle Endures

As this book has taken shape during the last decade, the world has moved apace, frighteningly so. As the world recovered from the COVID-19 pandemic and a global racial reckoning fomented from my hometown, I felt a sense of both urgency and paralysis revising this book. Working through personal and global tumult, I did not anticipate that the vociferous collective movement and protests referred to as *Aragalaya* ("struggle" or "public uprising" in Sinhala) that took shape and power in the spring and summer of 2022 in Sri Lanka would literally bring political power—a power viewed critically in this book—to a standstill. While the book concludes with the concerning dynamics of pandemic nationalism, I begin with the Aragalaya as exemplary of political possibility. The Aragalaya incited new, organic collective political hopes and spaces, cutting across ethnic, class, gender, and racial lines while also reproducing familiar forms of militarized and violent state repression. Comprehensive outlines of events and on-the-ground observations and reports are available elsewhere; here, a brief synopsis will have to suffice.[1]

In 2022, Sri Lanka was facing its worst economic crisis since independence in 1948, leading to ballooning inflation, electricity cuts, and shortages of basic necessities. A critical intersection of "external shocks" such as the COVID-19 pandemic and Russia's invasion of Ukraine combined with then president Gotabaya Rajapaksa's fiscal mismanagement and poor policy decisions to catalyze a major economic downturn.[2] This crisis was felt by millions of Sri Lankans, who heard the promise of Rajapaksa's national vision, "Vistas of Prosperity and Splendour," but never saw it materialize. Running out of foreign currency reserves, the government could no longer import essentials like fuel, medicine, and food, a situation that created massive shortages, seven-hour-long electricity cuts, and long fuel queues in which people actually died waiting. This came on top of the already steep rise in cost of living due to skyrocketing inflation. Despite the spike of infrastructural development and economic jumps after the end of the war in 2009, this economic turmoil was years in the making. The so-called peace dividend did not pay out. In postwar disaster nationalism and capitalism, undercurrents of Sinhala Buddhist nationalism, the concentration of executive power, increased military spending, and corruption and patronage politics led to neither development nor security. Sri Lanka's economic downfall has deeper roots, too, in economic liberalization as well as in the concentration of power to the executive presidency in 1978.[3]

Protests against the Rajapaksa administration started gaining momentum and numbers in March 2022, leading to a nonviolent occupation of Galle Face Green, the *GotaGoGama* ("Got[abaya] go village" in Sinhala) in April. Across social media, the hashtag #gotagohome also became prominent during this time. Galle Face Green in Colombo was also the site of the war victory parade I discuss in chapter 1. Many people came together during the occupation, ringing in the Sinhala Tamil New Year in April and breaking fast during Ramadan. Through the Aragalaya, people found a shared space in which to voice their concerns beyond economic woes. Years of authoritarian and repressive state violence controlled the narratives and memories of war and national security, but the ghosts that continue to haunt postwar Sri Lanka also materialized (Fedricks et al. 2023, 31).[4] Protests continued regularly beyond Galle Face Green. In May 2022, Prime Minister Mahinda Rajapaksa resigned. Of course the government responded in the way most familiar to them: through securitization, police force, and violence, yet *GotaGoGama* remained steadfast. I watched breathlessly on social media as protesters eventually overtook President Gotabaya Rajapaksa's home in July 2022. The president fled the country, tendering his resignation from Singapore shortly after. Longtime politician Ranil Wickremesinghe was installed as Rajapaksa's replacement—not exactly the regime change the Aragalaya desired. Echoing his predecessors by calling the protesters a "fascist threat to democracy," Wickremesinghe resorted to the Prevention of Terrorism Act and declared a state of emergency to "do whatever it takes" to stop the protesters. Arrests of students and protesters abound. Sri Lanka has taken on a loan from the International Monetary Fund and implemented various austerity measures. The status quo remains. Rice and fuel prices are only marginally lower. "What to do?" my friend in Sri Lanka responds to me with that familiar refrain over WhatsApp.

Still, it is possible to recognize the power of the Aragalaya—a collective force unimaginable and unprecedented, a new political space, demanding not just a new president or administration but a *system* change. The system in question is one that is critically examined in this book, a system of centralized power undergirded by Sinhala Buddhist nationalism and ennobled by the power of violence, which makes the Aragalaya's possibility for change, however fleeting, so extraordinary.

Acknowledgments

This book has been so long in the making. It is easy to begin this long list of gratitude, but it is difficult to know where to stop. I know there are names that I have unintentionally forgotten to add here.

First: the generosity of those who took the time to have conversations with me. Some of these conversations made it in this book, many did not, but all taught me about life and disaster in Sri Lanka. Most of these conversations could not have happened without Usha and Nitharshini, my amazing and fearless research assistants. They translated difficult conversations, hauled me around on the back of their scooters to wherever research whims took us, and always maintained good spirits, despite the gravity of our work. I had the most loving hosts with the Bartholmeusz family in Kalmunai. They treated me like their own sister, fed me delicious foods, while still giving me my own space and respecting my sometimes wonky research schedule. Muradh, a tireless and joyful leader of his community and coordinator of disaster mitigation and management, became a good friend. Many of our "work" meetings turned into long conversations and meals with his family and walks around his village. Fajru was one of the very first people I met out east, when I was struggling to put together a project early in graduate school. His generosity and critical concern for institutional forms of disaster management and development in the east were necessary and taught me about the ambivalences of doing disaster recovery work in Sri Lanka. His compassion was always for the people. As I was completing copyedits for the book, Fajru suddenly died. Fajru, I'm sorry I couldn't share the book with you in Sri

Lanka. May Allah grant you the highest level *Jannah*. Also gone too soon are brothers Lathan and Nalin, may they rest in power. Andrew Lucas and his treehouse were a home away from home. Omar Siddique was my Colombo refuge. He and Melissa Mandor were levity during a time of deep political turmoil and gravity in Sri Lanka. Sabrina Cader was welcome assistance in the National Archives. Nalayani Jayaram was a patient Tamil tutor (my own language shortcomings notwithstanding!) in Colombo who also generously shared with me her own family and life experiences.

I often joke that I have been sounding like a broken record for years, but this is a reflection of the many opportunities to give talks and presentations, all of which have helped me to think about the project in productive ways. Thanks to the following institutions: The Department of Anthropology at UC Irvine; the Department of Anthropology at Cornell University; the Department of Anthropology at the University of Tennessee, Knoxville; the Department of Anthropology at Oregon State University; the Department of Science and Technology Studies at Rensselaer Polytechnic Institute; the History and Philosophy of Science Workshop at the University of Chicago; the Center for South Asia at Brown University; Rutgers University's Symposium on Visual Culture in and out of Crisis; the Department of Anthropology and the Center for South Asia at the University of Washington; the South Asia Program at the University of Minnesota; the Department of Anthropology at Western Carolina State University; the Department of Science and Technology Policy at the Korea Advanced Institute for Science and Technology; the Department of Anthropology at Yonsei University; and the Khmer Studies and Social Work Program at the Royal University of Phnom Penh.

The research that formed the basis for this book was supported by a Fulbright Hays Doctoral Dissertation Research Fellowship and an American Institute for Sri Lankan Studies Postdoctoral Fellowship. The writing was supported by a UC Humanities Institute Dissertation Writing Grant, a Wenner-Gren Hunt Postdoctoral Writing Fellowship, and a fellowship from the National Endowment for the Humanities. Any findings, conclusions, or recommendations expressed in the book do not necessarily represent those of the National Endowment for the Humanities. Being at a teaching-focused liberal arts college, I recognize that there are fewer resources to support this kind of publishing, so I am grateful for a generous Professional Development Grant from the Faculty Life Committee at St. Olaf College and indexing funds from my Associate Dean dream team of Susan Smalling and Jennifer Kwon Dobbs.

Parts of this book have appeared in other places. Early thinking on disaster and insecurity appeared in "A Safer Sri Lanka? Technology, Security and Preparedness" in *Tsunami in a Time of War: Aid, Activism and Reconstruction in Sri Lanka and Aceh* (edited by Malathi de Alwis and Eva-Lotta Hedman) and "After Disasters: Emergences of Insecurity in Sri Lanka" in *Dynamics of Disaster: Lessons on Risk, Response and Recovery* (edited by Barbara Allen and Rachel Dowty). Much of chapter 2 was published as "Anticipatory States: Tsunami, War, and Insecurity in Sri Lanka" in *Cultural Anthropology*. Parts of the book appear in "Infrastructures of Feeling: The Sense and Governance of Disasters in Sri Lanka" in *Disastrous Times: Beyond Environmental Crisis in Urbanizing Asia* (edited by Eli Elinoff and Tyson Vaughan).

I first spent time in Sri Lanka on a Fulbright Fellowship, with not much experience and a lot to learn, prior to starting graduate school. I am grateful in particular to Malathi de Alwis, who generously gave a chance to my naive younger self. I regret that I was not able to share this book with her before her untimely death. I was fortunate to also have an amazing intellectual community at the University of California, Davis, guiding my research. My adviser, Smriti Srinivas, while always ready with readings, advice, and ideas, made it clear from day one that my project was mine. I treasure her wit, creativity, and willingness to think with me—she taught me very explicitly that the relationship between adviser and advisee was one of mutual intellectual exchange. My committee was a warm circle of generosity, insight, and care. Tim Choy first inspired me with his beautiful prose, and today he still provides gentle nudges, clever advice, and corny jokes (innovation!). Joe Dumit was the master of listening and repeating back to me what I had said in a way that made more sense while insisting that it was all my idea. A feat indeed! Alan Klima assured me during my first year of graduate school that I would "make it." He has been many academics' writing guru, including mine. Alan was instrumental in helping me organize the edits for the book while also reminding me that the book needed to be out in the world. They have all taught me what it means to be a generous scholar. I am also thankful for the opportunity to have learned from Marisol de la Cadena, Suad Joseph, Cristiana Giordano, and Suzana Sawyer.

Kim Fortun deserves special mention; as both mentor and friend, her unwavering support has been immeasurable. Among many intellectual and theoretical lessons, she has modeled the value of creating communities of care and making meaningful structural change. Thank you for including me in these communities, including your family. Gratitude also to Mike Fortun for his care, wit, humor, and outstanding cooking. Special shout-out to Kora and Lena!

UC Davis's program in sociocultural anthropology was special. After leaving I became especially aware of what a unique experience of camaraderie it had been. I recall conversations, food, inspiration, and commiseration with fond nostalgia: Jenn Aengst, Adam Brown, Madeline Otis Campbell, Jake Culbertson, Nicholas D'Avella, Jonathan Echeverri, Stefanie Graeter, Bascom Guffin, Chris Kortright, Jieun Lee, Tim Murphy, Jorge Nuñez, Charles Pearson, Rima Praspaliauskiene, Michelle Stewart, Lauren Szczeny-Pumarada, Leah Wiste, and Adrian Yen.

My anthropology world is buttressed by the following lovely people who have always had time for a conversation and a reassuring hug: Andrea Ballastero, Lee Douglas, Radhika Govindrajan, Stuart Kirsch aka "Prof. K," Nidhi Mahajan, Juno Parreñas, Noah Tamarkin, Sharika Thiranagama. Thank you to the following who are sources of both intellectual and personal sustenance: Nicholas D'Avella's friendship has been immensely important beyond the trials and tribulations of graduate school. Jerry Zee's energy and creativity are infectious. Stefanie Graeter, in addition to things anthropology, showed me how to be a dog mom. Jenna Grant: your mom and my dad are hopefully somewhere together toasting our respective books. Sorry mine came out after yours, stymieing our joint party. I am always in awe of Mythri Jegathesan's generosity. Critical disaster scholarship has grown so much as this book has developed, and Scott Knowles and Kim Fortun have been critical to my development as a disaster scholar. Fellow disaster scholars Megan Finn and Beth Reddy enthusiastically offered to read parts of the manuscript. The scholarly community in Sri Lanka is small yet mighty. Thanks to Eva Ambos, Malathi de Alwis, Nalika Gajaweera, V. V. (Sugi) Ganeshananthan, Mythri Jegathesan, Neena Mahadev, Dennis McGilvray, Nihal Perera, Alessandra Radicati, Ben Schonthal, Jim Sykes (personal communication!), and Sharika Thiranagama for helping me to feel a part of it. Amarakeerthi Liyanage was a wonderful summer FLAS Sinhala teacher. He confirmed a few translations for me in the book. Sharika, Mythri, and Sugi provided much-needed rallying as I finished the revisions for the book.

The Writing with Light Collective has taught me so much about shared visions (literal and figurative), and it is a genuine privilege to know and work with Craig Campbell, Lee Douglas, Arjun Shankar, and Mark Westmoreland. Thanks especially to Michelle Stewart, who early on had the idea of creating *Cultural Anthropology*'s Photo Essay Initiative and roped me into its development. Writing with Light has transformed that initiative into a new ambitious project. My co-curators gently encouraged me to share my images that are interspersed throughout the book.

My first position after graduate school was at Cornell University: I had two incredible years as a Mellon Postdoctoral Fellow at the Society for the Humanities and in the Department of Science and Technology Studies. The conversations and generative insights of colleagues at the Society during the respective thematic years of "Risk" and "Occupation" were challenging and inspiring. I am grateful to Bernardo Brown, Julia Chang, Rishad Chaudhury, Lorenzo Fabbri, Anna Watkins Fisher, Patty Keller, Nidhi Mahajan, Annie McClanahan, J. Lorenzo Perillo, and Antoine Traisnel for their scholarship and friendship while at Cornell. Anne Blackburn welcomed me to the South Asia Program and hosted a number of Sri Lankan scholars as director—an unparalleled experience for me most certainly. In STS, I thank Rachel Prentice, Sara Pritchard, and Suman Seth for their support and friendly advice. I was also lucky enough to become a part of a writing group of powerful women who read and commented on early pieces of this book as well and helped see my first peer-reviewed journal publication through: Maria Fernandez, Durba Ghosh, Rachel Prentice, Sara Pritchard, Kathleen Vogel, and Marina Welker.

The University of Tennessee's Program on Disasters, Displacement, and Human Rights gave me the opportunity to teach and work with some amazing students, like Christine Bailey and Carolina De La Torre Ugarte. My friendships with Gerard Cohen-Vrignaud, Pat Grzanka, Jasmine Fox, and Joe Miles helped me to more appreciate life in Knoxville.

The Department of Sociology and Anthropology has been such a wonderful home for me at St. Olaf College. The collegiality and care of Ibtesam Al Atiyat, Chris Chiappari, Andrea Conger, Marc David, David Schalliol, Ryan Shepherd, and Tom Williamson have carried me through tenure and beyond. Jennifer Kwon Dobbs has been a lifeline. Joanne Quimby has been my main cohort buddy since August 2016! The programs in Race and Ethnic Studies and Environmental Studies have also been warm homes outside my department. Melissa Flynn Hager in the grants office guided me through several successful grants. Lori Middeldorp and Jessica VanZuilen have made the hallways of Holland Hall more joyful. Sara Dale, GIS whiz, generously created the maps that appear in this book during her own free time and also taught me about using accessible colors. I have had the incredible luck of working with many amazing students, reminding me that the work anthropologists do can matter. Anna Clements, my research assistant for a summer, helped me straighten out my citations and bibliography. Kgomotso Magagula's enthusiasm for life and anthropology is a reminder of the joys and privilege of teaching. Or Pansky's tireless efforts against racism and Zionist settler colonialism continue to educate me, especially as, at this moment of writing, Israel carries out brutal genocidal violence against Palestinians.

During my sabbatical in Korea, the graduate program in Science and Technology Policy at the Korea Advanced Institute for Science and Technology (KAIST) was a friendly and exciting home for a semester. I am grateful to Scott Knowles, tireless disaster scholar, who has encouragingly insisted that what I have to say matters. My graduate seminar, "STS in Endangered Worlds," kindly read a draft of the book; their generous readings gave me a final boost of confidence to submit the revisions.

Minneapolis has been my home for the last eight years, and I have been fortunate to become part of an amazing friend group. With deep gratitude to Lorenzo Fabbri and Jennie Row (and Emma/Momo and now Livia Juni!) for their support, especially during some tough COVID times. I cherish the countless friendly meals, libations, and laughter of the LHH. Siri Suh's presence in Minneapolis is dearly missed. Sugi Ganeshananthan, by a stroke of luck—or fate, really—is the best neighbor I could have asked for when I first moved here. Her beautiful, prize-winning novel *Brotherless Night*, about the war in Sri Lanka and the experience of Tamils in Jaffna, models that sharing life stories and histories beautifully is possible. Sugi offered to write the short epigraphs that appear in the chapter interludes of the book; my chapters can only aspire toward such elegance.

The ocean, the desert, the forests, and their bounty have also been my teachers. There are many friends outside of academic life who have also provided sustenance, encouragement, and respite. They are too numerous to name, spread across the different geographies of my life: Bay Area, NYC, Portland, and Pomona crews. Amanda Gehrke has been a steadfastly warm and generous friend for decades; thank you for all the golden times.

My editor, Ken Wissoker, saw something in this project before I even really knew what it was going to be, and I am grateful for his confidence and patience. Kate Mullen was valiant in her efforts to get all the pieces of this book together. Thanks to Liz Smith and her team for their diligent copyediting skills. Cathy Hannabach at Ideas on Fire did stellar indexing work.

Before any of this research began, of course, my family provided me with every opportunity to explore my curiosity about the world. My parents, Joon and Hyung Choi, never begrudged me any educational experience growing up, and without their support, I would not have been able to pursue this research or anthropology more broadly. I dedicate this book to them. My *appa* died in 2014, long before this book came out. It took me some time to feel like I could get back to this project after that. Even though we profoundly disagreed on things ("What *are* they teaching you in grad school?!"), I know he would be proud. And my *umma* continues to be a never-ending source of love and support. My brother Brian and sister-in-law Margot have always been a solid and stable support system, and I love being *gohmo* Viv to their charming and rambunctious twin boys Archie and Theo. My sister Lillian has been a best friend and unwavering foundation throughout my life; I am also blessed to be the best *eemo* I can be to her son Auggie, who is a joyful miracle. Ross Chergosky has made Minneapolis more my home; he is so much love and laughter, unfailing support and generosity. Loves you. I look forward to continuing our life adventures. The best "present" to myself before moving to Minneapolis was my faithful companion species, Ollie. I am comforted by his snores as I write these acknowledgments.

The sea pronounces something, over and over,
in a hoarse whisper; I cannot quite make it out.
But God knows I have tried.

Annie Dillard, *Teaching a Stone to Talk*

Tsunami-damaged home on the east coast of Sri Lanka. Photo by author.

Introduction

WHEN DOES ONE DISASTER END AND ANOTHER BEGIN?

Mufeetha heard a deep and disturbingly quiet roar and looked up at the sky, which filled her with the fear of Allah. She thought at first she was looking at a sky blackened with crows. Later, she—along with many other Sri Lankans—would learn what to call that dark wall of water: tsunami. For a flickering moment, she thought of the valuables at home, her gold jewelry especially, but there was no time. With black waves lurching behind her, she managed to grab her two children and head inland. She was lucky to have escaped with her family (her husband was working abroad, as so many in Sri Lanka do), though her house was damaged. Many others in her community were not so fortunate: the eastern coast of Sri Lanka, the region of focus for this book and the location of Mufeetha's village, suffered more deaths and destruction than other tsunami-affected parts of the island.

On December 26, 2004, a massive megathrust earthquake with a magnitude of 9.1 rocked the Sunda Trench off the coast of Indonesia (Lay et al. 2005). The earthquake was registered as the third-strongest in recorded seismological history, with the longest duration of faulting—between eight and ten minutes—ever observed (Park et al. 2005). It was so strong it caused the entire planet to wobble, vibrating as much as half an inch off its axis. It also tore open a gash in the earth between 720 and 780 miles (1,200–1,300 km) long (National Science Foundation 2005). The subduction of the India Plate beneath the Burma Plate also triggered a series of deadly tsunamis along the coastlines of many landmasses in the Indian Ocean. At the time, the "Boxing Day Tsunami" was a natural disaster of unprecedented magnitude. More than 230,000 people died in fourteen countries. Aceh, Indonesia, located nearest to the epicenter of the earthquake, was the most devastated region, with over 160,000 deaths. Sri Lanka was the second-most devastated country, with over 35,000 dead or missing and over 500,000 displaced (Government of Sri Lanka 2005), prompting global sympathy and unprecedented levels of international aid and response (Korf 2006a; Telford and Cosgrave 2007) (see map I.1).

The black waters of the tsunami struck an already war-torn shoreline in Sri Lanka. In 2004, Sri Lanka was engaged in what was then Asia's longest-running civil war, in which the Sri Lankan government had been battling the militant separatist group the Liberation Tigers of Tamil Eelam ("LTTE" or "Tigers") for over two decades.

Disaster Nationalism traces the politics of tsunami reconstruction as they unfolded upon an already scarred social and political landscape in Sri Lanka. Given the island's decades of strife engendered by ardent, exclusionary, and oftentimes violent nationalisms, I argue that the tsunami's devastation and the techniques of disaster management and reconstruction that followed created opportunities for new modes of statecraft, national restructuring, and militarization. With the Sri Lankan government functioning as the obligatory passage point (Callon 1986) through which disaster management practices and reconstruction programs were conceived and executed, post-tsunami reconstruction efforts such as national disaster warning systems, coastal no-build buffer zones, and new housing schemes served as material, physical, and ideological nation-building projects. These efforts legitimated new forms of population and territorial management as well as—most significantly, as this book will detail—the government's aggressive approach to the war and terrorism (see Deleuze 1992; Foucault 2007; Lakoff and Collier 2015; Ong and Collier 2005). Based on eighteen months of fieldwork spanning from 2008 to 2017, *Disaster Nationalism* follows these national disaster management projects after the tsunami and through the end of the civil war in May 2009.

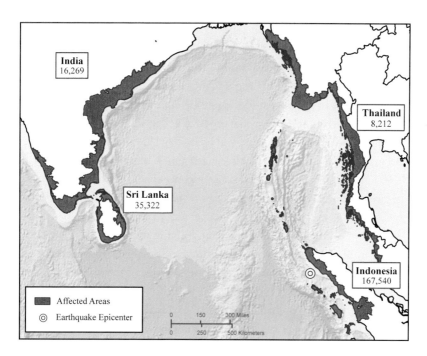

MAP I.1. 2004 Tsunami: Most affected countries. Total count of missing or dead as a result of the 2004 tsunami in Indonesia. Source: International Tsunami Information Center—UNESCO. Credit: Sara Dale.

As the organizing framework and heuristic of the book, I define disaster nationalism as both the process and mechanisms of state power that utilize disasters to produce, legitimize, and entrench national ideologies. As Naomi Klein's conception of "disaster capitalism" points to the way disasters open the door for corporate and free-market reorganization, disaster nationalism highlights a different but related phenomenon that took shape in Sri Lanka amid the ruins of both the civil war and the Indian Ocean tsunami. Where decades of war and competing virulent nationalisms had normalized militarization as a structuring force in social and political life, disaster management as a newly established institution after the tsunami also cohered with and empowered the Sri Lankan government's military goal to eliminate the LTTE.[1] The war was not merely the social context in which the tsunami played out; rather the tsunami reinforced an existing militarized logic employed to manage uncontrollable threats, including war and terrorism. Disaster management worked as a legitimate institutional framework categorizing both tsunamis and terrorism as imminent disaster risks, in turn sanctioning the state's goals to protect and uphold existing exclusionary majoritarian Sinhala Buddhist nationalist ideologies. By examining the militarization of disaster management, I show how the Sri Lankan state propagates the fantasy of an always-at-risk nation. And by tracing the mechanisms of disaster nationalism, I illustrate how disaster management also became a mechanism of national securitization, reproducing ongoing conditions of insecurity and precarity for minority Tamil and Muslim communities in disaster-affected areas of the island. This palpable lack of social and political change in the years *after* the tsunami and *after* the end of the war evinces the enduring disaster that is state-sponsored nationalism in Sri Lanka: "That there should be no difference between disaster and none at all: this is the disaster" (Smock in Blanchot 1995, xiv–xv).

Disaster serves as both empirical focus and analytic in this book. Empirically, disasters challenge and exhaust existing idioms, epistemologies, methods, and politics (Fortun 2001, 2012; Fortun et al. 2017). As totalizing and multifaceted events and processes of scale, disasters are not easily contained, materially, temporally, or categorically. They are also prisms, allowing us to see novel forms of social and political response (see Guggenheim 2014; Tironi 2014). They lay bare social and political instabilities while exacerbating and creating new ones. Each disaster shifts existing ontologies of disasters, the intersections of the material and social worlds, challenging the horizons of what is possible, what is imaginable (Morimoto 2012; Oliver-Smith 2002). While traditionally war has been absent from disaster categories outlined by the United Nations Office for Disaster Risk Reduction and the Sendai Framework, as Kenneth Hewitt (2021) urges, the impacts of war demand the attention of critical disaster studies. The intertwined disasters of tsunami and civil war in Sri Lanka forced me to reckon with their varied, complex, and layered elements, dimensions, and articulations. In turn, disaster also became a productive analytic.[2]

In earlier stages of research, I was concerned that "disaster" should be a category used judiciously. Sri Lankan disaster governance, and disaster risk reduction and management more generally, by employing the term expansively, had created opportune conditions for power. I worried, too, that "disaster" might lose its significance or become banal through overuse. But I was challenged to consider, on the contrary, what it might mean to consider more phenomena in the world as disasters.³ Why shouldn't other forms of injury, destruction, and violence be given the same attention that spectacular disastrous events are? My approach to disaster as an analytic, then, is less invested in quibbling over what it is, what counts as a disaster, and the word's lexical origins or confusions and conflations with related terms such as crisis and catastrophe (see, for example, Barrios 2016; Quarantelli 1998; Tierney, Lindell, and Perry 2001), and more about following its contextual articulations and movements, considering disaster as both an object of concern (and care) and a process. Accordingly, in this book I follow disaster's articulations as an empirical force, materially, emotionally, discursively, and institutionally, as it is mobilized and militarized by the Sri Lankan government and, crucially, as it is experienced by Sri Lankans. It is precisely the power of conflating terrorism and natural disaster *as* disaster that this book problematizes. I build on anthropological work calling for systemic analyses of disaster as process (Button and Schuller 2016; Oliver-Smith 2002) by considering historical and political circumstances not just as prefigurations of risk and vulnerability but also *as disasters*. State-sponsored nationalism *as* enduring disaster in Sri Lanka reveals the temporal and political stakes, structures, and experiences of disaster nationalism.⁴

As an analytical lens, then, disasters and their unfoldings and management draw out how different kinds of political systems—racial, colonial, economic, technological—intersect. Since the tsunami and earthquake in 2004, disasters such as Japan's "Triple Disaster" of March 2011 (Dudden 2012; Pritchard 2010; see also Allison 2013), Hurricane Katrina (Adams 2013; Carter 2019), Hurricanes Maria and Ida (Bonilla 2020; Lloréns 2021) and earthquakes in Nepal (Shneiderman et al. 2023; Seale-Feldman 2020; Warner, Hindman, and Snellinger 2015) and Haiti (Beckett 2020; Farmer 2011; Schuller 2016) show how disasters have shaped and will continue to shape life and politics in the Anthropocene.⁵ Moreover, as emergencies increasingly justify exceptional modes of militarized humanitarianism as a form of global governance (Benton 2017; Fassin and Pandolfi 2010) and as a counterinsurgency tactic (Bhan 2014; on militarized humanitarianism as "warfare," see Zia 2019), the need for examining these complex intersections and how they are experienced remains ever urgent. *Disaster Nationalism* offers both a method and a theory to examine the implications of new modes of risk and disaster infrastructures and technological fixes (see Fisch 2022; Reddy 2023) as preemptive approaches to disasters become more salient in contemporary security-scapes (Gusterson 2004) of anti-terrorism (Anderson 2010; Masco 2014), climate security (Cons 2018), and, especially in this moment of writing, pandemics (Keck 2020; Porter 2019).[6]

NATIONALISM AND MILITARIZATION IN SRI LANKA: A BRIEF HISTORY

To historicize the intersection of nationalism and militarization in Sri Lanka, this section provides an abbreviated outline of the civil war (with chapter 1 providing a more tailored account of institutional precursors to the Sri Lankan government's Disaster Management Act). In short, the war, which spanned from 1983 to 2009, was waged between the Government of Sri Lanka and the LTTE, a militant rebel group fighting for what they believed to be their rightful homeland on the island. The war did not break out suddenly, but rather came to a head in 1983. As I detail below, following independence from Great Britain in 1948, Sinhala Buddhist majoritarian nationalist policies and social projects increasingly marginalized and alienated Tamil minorities, leading to a secessionist movement and ultimately the brutal emergence of the LTTE as the self-proclaimed representative of Tamils and Tamil Eelam in Sri Lanka—and decades of civil war.

First, to give a sense of Sri Lanka's heterogeneity, some demographics. The majority ethnic group at 75 percent is Sinhalese; Tamil populations central to the conflict and the notion of the imagined monoethnic Tamil Eelam (see map 1.2) make up approximately 11 percent of the population (a number that has fluctuated down due to migration and deaths from war), and Muslims (classified as both an ethnic and religious minority) make up approximately 9 percent. Malaiyaha or "Hill Country" Tamils, descendants of South Indian plantation laborers who migrated during British rule, are another Tamil-speaking minority and make up 4 percent of the population. Other minorities such as Portuguese and Dutch Burghers and the aboriginal Veddas make up the remainder of Sri Lanka's population. Tamil and Sinhala are also languages, though Muslims, mainly in the Northern and Eastern Provinces, also speak Tamil. Many other minorities speak Sinhala, Tamil, and English. Sinhalese are predominantly Buddhist (approximately 70 percent), though some are Christian. While Tamils are mostly Hindu (approximately 13 percent), a small subset are also Christian.[7]

While in shorthand the civil war is often characterized as an "ethnic" one, ethnicity fails to capture the complexities of the conflict. The majoritarian nationalism of Sinhala Buddhism is not just about religious and ethnic claims but also about how these claims are at once political, cultural, and economic (Hewage 2014; Kadirgamar 2013; Venugopal 2011). Relatedly, I shift the frame to illuminate the consequences of state formation (see Bastian 1999) in which Sinhala Buddhist nationalism is central, yet not all-encompassing. The foundations of state-sponsored projects of exclusionary nationalism, decades of an almost continuous state of emergency, and legally sanctioned counterterrorism measures and violence all point to how disaster can be instrumentalized by the state. Disaster nationalism draws attention to the mechanisms of disaster management that merge with state, military, and Sinhala Buddhist nationalist ideologies.

Prior to the "disaster" (Manor 1984) of Black July in 1983 and the official "beginning" of the Sri Lankan civil war, tensions had long been mounting between the majoritarian Sinhalese Buddhist government and minority Tamils. Though ardent nationalisms have pitted Sinhalese and Tamils against each other since time immemorial, Sri Lanka's ethnic identities and affiliations are rather more recent historical developments, made politically meaningful first through British colonial governance and later as Sri Lanka (then Ceylon) developed as a postcolonial nation-state.

While the British did not "invent" ethnicity (Thiranagama 2011), they did foster notions of "racial" difference that attained increasing significance through political structures. In the gradual centralization of state power after independence, a new and powerful Sinhala nationalist consciousness alleged that the majority community—the Sinhalese—had been exploited by colonial rule, which had also given undue influence to minority groups including Tamils, Muslims, and Christians (Jayawardena 2003; Tambiah 1986; Thiranagama 2011). The idea of Buddhism as a response to colonialism "laid the groundwork" for both Sinhala Buddhist identity and Sinhala Buddhist nationalism (Gajaweera 2015). This hegemonic Sinhalese national consciousness resulted in discriminatory policies that I outline in detail in chapter 1.[8] The Constitution of 1972 changed the nation's name from Ceylon to Sri Lanka, removed protections for minorities, and enshrined Buddhism as the official state religion, illustrating the growing power of a Sinhala *and* Buddhist national identity. The exclusionary power of these various ethno-religious political developments is evidenced by the numerous anti-Tamil riots in 1956, 1958, 1977, 1981, and 1983.

Through the 1970s, intense state-sponsored Sinhalization led to the social tensions which gave rise to Tamil political movements (Rajasingham-Senanayake 1999; Venugopal 2018). Facing increased marginalization and racism in Sri Lankan politics, the major opposition party, Tamil United Liberation Front (TULF), emerged, and with it the notion of an independent state of *Tamil Eelam* (see map I.2). Composed mainly of middle-class and upper-caste educated Tamil gentlemen, TULF was committed to a nonviolent solution.[9] Despite this, the notion of an independent Eelam roused suspicion among the Sinhala-dominated government. This suspicion was heightened by the growing presence of a militant group of young, lower-caste Tamil men—the Tamil New Tigers—who had begun committing acts of robbery and killing Sinhalese police officers. As Ahilan Kadirgamar (2020) recounts, this armed presence, combined with the 1977 anti-Tamil riots after elections and growing state repression, created more patterns of violence on the island.

By July 1983, then, social tensions had been mounting. During this period, the New Tigers continued to gain control of the Eelam movement, claiming to be the sole voice of Sri Lankan Tamils as the Liberation Tigers of Tamil Eelam (LTTE) and fighting for what they saw as their rightful monoethnic imagined homeland of the northern and eastern regions of the island.[10] The violence against Tamils that police were committing with impunity in Jaffna stoked the flames of the Tiger militants, eventually leading to retaliatory actions by both parties. The final act that led to the violent riots of July 1983 was the killing of a convoy of Sinhalese policemen by the Tigers in their northern stronghold of Jaffna. Riots against Tamils ensued in areas all over Sri Lanka. The most striking aspect of the 1983 anti-Tamil riots was the state-sponsored nature of their disorder and violence: police, if not committing violent acts themselves, stood by and watched as they unfolded (Jeganathan 2000; Manor 1984).

After 1983, the war would take several twists and turns, with many failed attempts to come to a peace accord or resolution. These various parts of the civil war are divided into four phases: Eelam War I, beginning in July 1983 and ending in 1987; Eelam War II (1987–1993); Eelam III (1994–2001); and Eelam IV (2005–2009). These phases reflect periods of intense hostilities and fighting.

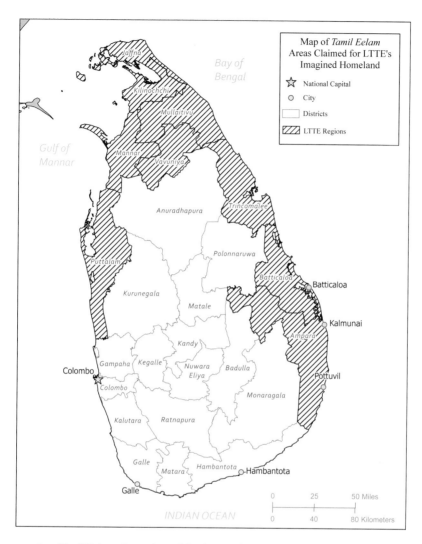

MAP I.2. Tamil Eelam: Areas claimed for the LTTE's imagined homeland.
Credit: Sara Dale.

Throughout the war, many atrocities would be committed. Both the government and the LTTE demonstrated their willingness to terrorize civilians. In addition to the LTTE, the Sri Lankan government's anti-terrorism efforts brutally put down the Janatha Vimukthi Peramuna (JVP or the People's Liberation Front) and their anti-state insurrectionist movements in 1971 and again in the late 1980s.[11] The LTTE, under the direction of their leader Velupillai Prabhakaran, forcibly recruited child soldiers, killed and assassinated vocal critics and fellow Tamils, and was credited with the invention of suicide bombing. Though the LTTE committed many acts of terrorism throughout the decades-long conflict, other actors—including the Sri Lankan state, the Indian Peacekeeping Force, the JVP, and other paramilitary groups[12]—also perpetrated violent overtures and extrajudicial killings. Militarization became embedded in social and political life, and in administrative culture, with state efforts to securitize spaces with soldiers and checkpoints (Pieris 2018).

Perhaps the most hopeful moment for a peace negotiation came in 2002, when the LTTE and the Sri Lankan government signed a ceasefire agreement mediated by the Norwegian government. However, the relative break in fighting would not last long; relations began to sour between the two warring parties again. In March of 2004, LTTE colonel Karuna Amman broke away from the Tigers, alleging that they had long ignored the needs and interests of eastern Tamils. Karuna's departure, and the defection of some 6,000 LTTE soldiers to the Sri Lankan Army, debilitated the LTTE's strength and reach in the east. Tensions between the LTTE and the Government of Sri Lanka grew increasingly heated until the tsunami crashed into Sri Lanka's already war-weary shorelines in December 2004, after which the final phase of the war, "Eelam IV," would begin. By that time, the death toll of the war had exceeded 60,000, with internally displaced populations sometimes as high as 800,000 (Le Billon and Waizenegger 2007). Tsunami reconstruction would be folded into the government's war campaign.

DISASTER MANAGEMENT: A SAFER SRI LANKA?

With thousands dead and missing and nearly three-quarters of Sri Lanka's coastline inundated, then president Chandrika Bandaranaike Kumaratunga and the government of Sri Lanka were chided locally and internationally for their slow and disorganized response to the crushing devastation. Heeding these criticisms, Kumaratunga declared a state of emergency, ushering in new institutionalized practices and legal frameworks around disasters and disaster management. For, if anything, the tsunami brought into sharp relief how *unprepared* Sri Lanka was for tsunamis and other natural disasters. Friends and administrators in Sri Lanka lamented to me that before 2004, they had no idea that tsunamis even existed. In response, the Sri Lankan government formed the Parliamentary Committee on Natural Disasters, whose mandate was to assess Sri Lanka's level of preparedness for such unexpected catastrophes. The culmination of this committee's work was Disaster Management Act No. 13, which "provides for a framework for disaster risk management in Sri Lanka and addresses disaster management (DM) holistically, leading to a policy shift from response based mechanisms to a proactive approach toward disaster risk management [DRM]" (Ministry of Environmental and Natural Resources 2007, 67; Ministry of Disaster Management 2005).[13] This proactive risk management approach was also presented in the committee's "Towards a Safer Sri Lanka: Road Map for Disaster Risk Management," in which risk and vulnerability assessments figured as key to creating a state of preparedness, as opposed to responsiveness, for whatever type of disaster may come.[14]

In this preparedness approach, risk figures as potential disaster and future threat to national security, and national threats are not limited to natural disasters but can also include health pandemics and terrorist attacks (Knowles 2013; Lakoff 2008; Lowe 2010; Massumi 2005). Under the purview of risk management, a preparedness rationale solicits new technical Band-Aids and infrastructure: warning systems, evacuation drills, event simulations, and overall attempts to increase government management and control. The impetus of such programs and collaborations is to invoke a continual state of readiness and maximum security of state territory: it is not a matter of *if* a disaster strikes, but a matter of *when*. The imminence of disaster is well-represented in the circular temporality of the disaster risk management framework: Mitigation → Preparedness → Response → Rehabilitation → Mitigation. In Sri Lanka, this shift toward preparedness was overseen by the National Disaster Management Centre (DMC). While specialized disaster agencies existed in Sri Lanka before the tsunami, there was no legal framework for disaster management and thus no holistic mechanism by which to coordinate it, and the DMC would fill this institutional gap.

The establishment of Sri Lanka's Disaster Management Act and "Roadmap for a Safer Sri Lanka" aligned with the Hyogo Declaration and Hyogo

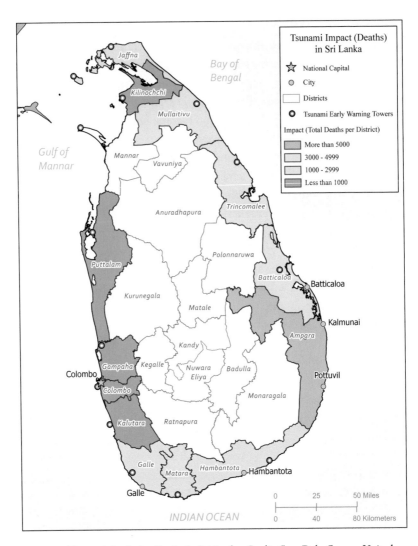

MAP I.3. Tsunami deaths by district in Sri Lanka. Credit: Sara Dale. Source: United Nations Office for the Coordination of Humanitarian Affairs.

Framework for Action 2005–2015, which was endorsed by the General Assembly of the United Nations at the World Conference on Disaster Reduction immediately following the tsunami in January 2005. The Hyogo Framework, while new, followed an existing ethic of prevention and conceptions of disaster risk management proposed by the United Nations International Strategy for Disaster Reduction (UNISDR) in 2000 (UNISDR 2007).[15] In the Hyogo Framework, disasters are limited to events "caused by hazards of natural origin and related environmental and technological hazards and risks" (UNISDR 2007). Alongside the tasks of rebuilding homes and villages and reestablishing livelihoods for people, new national disaster preparedness projects were also being implemented as post-tsunami projects with the support of the United Nations Development Program (UNDP), which aimed to work closely with the Sri Lankan government's development strategies. Sri Lanka's Disaster Management Act of 2005 defines disasters and risks more expansively. It categorizes disasters as "the actual or imminent occurrence of a *natural or man-made* event, which endangers or threatens to endanger the safety or health of any person or group of persons in Sri Lanka, or which destroys or damages or threatens to destroy or damage any property" (Ministry of Disaster Management 2005; my emphasis). These "natural or man-made" events include floods, landslides, industrial hazards, tsunami (seismic waves), earthquakes, air hazards, fire, epidemics, explosions, air raids, civil or internal strife, chemical accidents, radiological emergencies, oil spills, nuclear disasters, urban and forest fires, and coastal erosion.

The shock of the tsunami raised hopes among citizens and politicians alike that the destruction might lead to a unified effort to rebuild the nation. As unprecedented amounts of humanitarian aid poured into Sri Lanka, Chandrika Bandaranaike and the Government of Sri Lanka attempted to forge a joint aid-sharing mechanism, the Post-Tsunami Operation Management Structure (P-TOMS), between the government and the LTTE.[16] Compared to the southern coast, tsunami aid distribution lagged in the LTTE-controlled north and east; the P-TOMS was intended to move aid in more quickly.[17] Unfortunately, the P-TOMS stoked already existing social and political cleavages and fears (see also Hyndman 2007). The Tigers clamored to have as much control of the funds as possible. Muslims (based on a history of distrust and vulnerability and exclusion in broader conflict-related peace talks) and Sinhala-Buddhist nationalist hardliners decried the P-TOMS as giving into LTTE's secessionist and sovereign demands with the support of international donors (Keenan 2010; McGilvray and Raheem 2007).[18] Tsunami-affected Sinhalese in Southern Sri Lanka saw the P-TOMS as a "precursor to dividing the nation" (Gamburd 2013, 12), in addition to concerns that the aid would be misappropriated into LTTE war funds, rather than actually used for tsunami aid. Fears of "unethical conversions" by Christian humanitarian organizations further "threatened" Buddhist foundations on the island (Mahadev 2014; see also Korf et al. 2010).[19]

Though the P-TOMS was eventually signed in June 2005, the Sri Lankan Supreme Court issued a stay to block its implementation in July. So while the 2004 tsunami did provide a fleeting respite from the volatile relations between the LTTE and the Sri Lankan government, disputes over the P-TOMS seemed to cast the final blow to the already tenuous ceasefire (Uyangoda 2005).[20]

Just a month after blocking the P-TOMS, the Supreme Court also announced that President Bandaranaike's term would end in December 2005. In November 2005, Mahinda Rajapaksa was elected on a platform to end the war. He soon replaced the previously established Task Force to Rebuild the Nation (TAFREN) with the Reconstruction and Development Agency (RADA), claiming that reconstruction would be parallel to the peace process. By 2006, the LTTE and the government were engaged in retaliatory fighting. The Sri Lankan government took military advantage of Karuna Amman's breakaway from the LTTE and their weakened hold over the east. In 2007, the eastern region of Sri Lanka, once considered part of Tamil Eelam, was under the control of government forces. By January 2008, the Sri Lankan government had declared the already tenuous ceasefire officially null. Sri Lanka's roadmap for a safer Sri Lanka would continue along a path of war.

DISASTER NATIONALISM

When I arrived in Colombo to begin my fieldwork in 2008, President Mahinda Rajapaksa's aim was clear. "Mahinda Chinthana [Mahinda's Intention]: Vision for a New Sri Lanka" pledged to bring a swifter resolution to the war, with an agenda that "renounced separatism" and promised a national security policy that would prioritize the sovereign and territorial integrity of the island. This vision was articulated in billboards that decorated Colombo and its suburbs, like the one pictured in figure I.1, which depicts the government's objective for the year of its sixtieth anniversary of independence: "The Year for War."

FIGURE I.1. "2008: Year for War" billboard in Colombo. Photo by author.

The far-left image in figure I.1 demarcates in bright red the areas controlled by the LTTE in 2005 and reads, "areas under terrorists." The center image illustrates the "areas where terrorists are hiding at the end of 2007." In 2007, the government seized military control of the eastern bloc of Eelam, leaving only the LTTE stronghold of the north under the control of "terrorists." The last map on the right is a fully green, glowing Sri Lanka—a nation freed from the clutches of terrorism. The bottom reads: "Let us build a country where all nationalities can live in freedom." This billboard was displayed for Sri Lanka's Independence Day celebrations; at the top it says: "Let us celebrate sixty years of independence with pride." Though the government did not fulfill their goal of ending the war in 2008, they did finally proclaim victory over the Tigers on May 19, 2009.

This national projection of a "free" Sri Lanka worked in tandem with the government's disaster management strategy following the tsunami. Disaster nationalism—a militarized logic and practice of managing potential risks to the nation—justified the state's approach to natural disasters *and* terrorism. Consider an address given by Minister of Disaster Management and Human

Rights Mahinda Samarasinghe a few months after the war's dramatic end. In his speech at the Second Session of the Global Platform for Disaster Risk Reduction in Geneva, the minister discussed the successes and potential successes of a Disaster Risk Reduction (DRR) framework and of mainstreaming DRR in national development planning. He ended the address by shifting his focus from natural disaster to the recent end of Sri Lanka's "man-made" disaster:

> My country—Sri Lanka—has just overcome a human-made disaster of a magnitude unparalleled by any similar recent events elsewhere. We have overcome the scourge of terrorism that has beset our island nation for well over two decades.... The Government of President Mahinda Rajapaksa has taken on the task of reconstructing a nation which has suffered much in its efforts to reunite its people ensuring that all Sri Lankans are now able to lay claim to one, undivided territory as their common Motherland.
>
> All our efforts at renewal, rebuilding and resettlement, however will be put at risk if the cause factors of the conflict and terrorism are not addressed and our President has committed himself to evolving a home grown political response to those factors. *Borrowing from DRR methodology*, our political response will reduce the risk of a renewed human-made disaster, i.e. terrorism and conflict, through systematic efforts to analyse and manage the causal factors, evolve consensual responses and improved preparedness for adverse events. We do not for a moment think that Sri Lanka's national renewal will be quick or easy.
>
> There are ever-present threats that we must, and will, guard against, including the threats of new violence and destabilization. (Samarasinghe 2009; my emphasis)

According to this institutionalized logic, disasters—both natural and human-made—remain ever possible, even after their supposed ends.

This logic would be reiterated again just a month later. In July 2009, I attended a symposium on Disaster Risk Reduction and Climate Variability with Farood, a man from the east who was also working on disaster-related development work. Once again Minister Samarasinghe offered his words on disaster governance in Sri Lanka. He did not immediately discuss climate change or the tsunami, but instead addressed the situation of some 280,000 war-displaced Tamils in the north, interned in military-guarded camps. He approached the topic by way of his recent trip to L'Aquila, Italy, devastated by an earthquake, where he had "toured" the displacement camps. In his speech, he emphasized that the entire visit was escorted and vetted by the Italian government and, further, that no agency had unrestricted access. By sharing the story of this "escorted visit," he suggested that "free media access" would disrupt the "fragile area." My friend jabbed me in the side with his elbow: "Write this down!" he whispered loudly into my ear. Samarasinghe stressed that it was unreasonable for international aid agencies to ask for unrestricted access, as "unrestricted access is contrary to the national framework." International assistance is welcome, the minister reiterated, as long as it falls within the national framework of an "independent nation." "You see," my friend told me later, "they cannot resist talking about war when it comes to disaster." Echoing the billboard discussed earlier, in the Sri Lankan state's fantasy production (Aretxaga 2003)—the reality of a purified nation—terrorism and nature are cast into the same category of an "other" or imminent "threat" that must be continually managed.

My term *disaster nationalism* builds on Naomi Klein's (2005) notion of disaster capitalism. For Klein, the "shock" of disasters including wars, natural disasters, coups, and terrorist attacks invites the conditions for advancing neoliberal economic reforms and restructuring under the guise of democracy building. Key to her analysis is not just the possibility of profiting from disaster but also the ways in which disasters and moments of shock and terror are exploited to impose radical social and economic engineering. In Sri Lanka, rebuilding efforts favored and drove profit-minded projects, in which strategic and predatory land grabs from vulnerable communities were used to facilitate lucrative tourist development (Wright, Kelman, and Dodds 2021). Along the southern coast, where the "golden wave" of humanitarian aid was competitive and steeped in patrimonial politics and neoliberal logics (Gamburd 2013; Korf et al. 2010; Stirrat 2006), concessions to the one-hundred-meter no-build buffer zone were made for hotel construction and tourist development. Meanwhile, those living in buffer zones, often fishermen whose livelihoods were at stake, were not allowed to rebuild their homes (see Gunewardena 2010). Along the eastern coast, some areas such as Arugam Bay were also being primed for tourist development. Further north, in Ampara District, where I worked, the creation of buffer zones exacerbated land and housing shortages and made reconstruction slow and difficult, but because there was hardly any tourist infrastructure to rebuild, it was not a priority.

Disaster capitalism has gained traction and drawn critical attention to the aftermaths of disaster, including other tsunami-affected regions such as Thailand (Cohen 2012; Rigg et al. 2008) and Southern India (Swamy 2021). However, as Schuller and Maldonado (2016) note, this theory or conceptual framing does not account for the long-term historical, racial, and colonial structures that create the conditions for predatory and accelerated forms of dispossession and development (see also Bonilla 2020). Disaster nationalism is certainly inspired by Klein to explore the ways that disaster opens the door for political restructuring in Sri Lanka, and also adds to it by considering the broader historical, ethnic, social, and political—in addition to the economic—foundations that enabled national and militaristic restructuring in Sri Lanka after the tsunami.

Disaster nationalism, then, does not preclude modes of disaster capitalism or work outside of capitalism. Rather, disaster capitalism, and neoliberal forces more generally, are also nationalist projects that resonate together amid state practices of security (Amoore 2013). As with economic development projects in the 1970s, post-tsunami and postwar development were not just "compatible with but reincarnated" ancient indigenous, Sinhala Buddhist national culture (Tennekoon 1988, 297; see also Jazeel 2013; Venugopal 2011). At the aforementioned disaster symposium, Minister Samarasinghe acknowledged that disasters, whether human and natural, could be impediments to national development and economic growth. Referring to the end of the war, the minister suggested that conflict resolution also required climate mitigation strategies. After all, he explained, Sri Lanka had been given another chance to live in "unity." These remarks aligned with rhetoric justifying the government's war campaign: defeating the Tigers would allegedly yield "peacetime dividends" for the island. Tracking with forms of disaster capitalism already in place after the tsunami, the state viewed the island's war-torn northeastern regions as blank slates for development and economic reconstruction "compatible" with Sinhala Buddhist nationalist power; in these areas, the Sri Lankan Army colonized newly "freed" land to build beachside resorts and hotels and to erect war victory monuments (see Buultjens, Ratnayake, and Gnanapala 2016; Hyndman and Amarasingam 2014).

As Benjamin Schonthal notes, "configurations" of Buddhist nationalism in Sri Lanka have been organized and motivated since at least the 1940s by the perception that Buddhism in Sri Lanka is under threat. Tracing three different Buddhist national movements, he stresses that while the movements may seem distinct, they are, in short, "new variation[s] of [this] older discursive template" (Schonthal 2016, 98). These are neither inevitable nor identical movements and histories. What is emblematic of these nationalist movements—and in governmental actions after the tsunami—are perceived threats to the territorial sovereignty and stability of a (by May 2009, victorious) Sinhala Buddhist nation. Thus, post-tsunami reconfiguration of Sri Lanka's majoritarian nationalist imaginary, though not necessarily new, was emboldened and galvanized by newly implemented logics and practices of disaster management. This was not the negotiated unification hoped for immediately after the disaster, but rather an ever-vigilant nation purified of enemies, threats, and risks through expulsion, death, and dispossession. Studying the tsunami and the civil war together *as disasters* illuminates how new technologies and techniques of statecraft also engender national fantasies. Disaster nationalism, then, highlights how these technologies and techniques define and use disaster to further national and political agendas, resonating with a familiar social and historical trajectory of violence propelled by ethnic and religious majoritarianism in Sri Lanka (Calhoun 2017; Simpson and Corbridge 2008; Simpson and de Alwis 2008).

This book traces the practices and technologies of disaster management as a mode of national securitization sustained by these national fantasies. As I will go on to show, rhetoric about preserving and upholding Sri Lanka's newfound freedom provided moral justification for increased militarism and securitization of areas of Sri Lanka after the tsunami and after the war, perpetuating insecurity and precarity for many Sri Lankans who have endured decades of disasters.

ETHNOGRAPHY IN TIMES OF DISASTER

The research for this book was primarily conducted in this context of post-tsunami reconstruction as it intersected with the end of the civil war, between 2008 and 2009, with follow-up visits in 2014, 2015, and 2017. I have spent time in Sri Lanka off and on since 2003, when I first worked with a Colombo-based Sri Lankan NGO immediately following the signing of the ceasefire agreement. I arrived at a hopeful moment, a hope that seemed to diminish each time I returned. When I started my fieldwork in earnest in 2008, I was aware of the crumbling ceasefire agreement and growing hostilities between the LTTE and the government; I never imagined that the war would

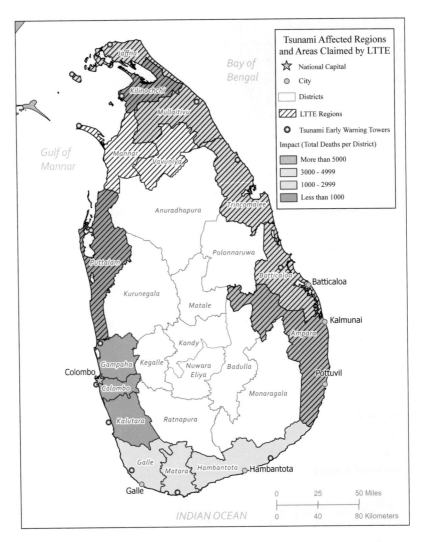

MAP I.4. Tsunami-affected areas and areas claimed by the LTTE. Credit: Sara Dale.

finally end, let alone that I would be in Sri Lanka during that final declaration. As I followed through with my post-tsunami reconstruction and disaster management research, the war shadowed everything, hovering over every place, every movement. No one was immune to or untouched by the conflict. How could it have been otherwise? The war was on everyone's mind, coming up in conversations everywhere. There were constant reminders that a war was happening in the country, even as the physical battles played out in a distant place, seemingly far away, unreachable without government approval (a situation discussed in more detail in chapter 4). It seemed that one could get used to all the billboards, handbills, news reports, text messages—even the armed police and checkpoints—only to have that sense of normalcy punctuated by, for example, a surprise air raid attack by the LTTE, a police roundup of the Tamil men working on a post-tsunami reconstruction project in the east, the accidental sounding of a tsunami warning siren, or rumors of an earthquake.

My fieldwork in Sri Lanka, which began in February 2008—just after President Rajapaksa declared the 2002 ceasefire null—was conducted over fifteen months during the Sri Lankan government's aggressive military campaign. I was still in the midst of my field research when the war came to a dramatic end on May 19, 2009. The beginning months of my field research were spent acclimating to the security situation, establishing connections in disaster management settings in Colombo, and figuring out how best to establish life and research in the east, where a tight security situation remained in place despite the region having been "liberated" by the government. Unfortunately, after six months, I had to return to the United States because my father was ill. I arrived in Sri Lanka in January 2009 for another nine months, nearly seven of which were spent on the east coast during the height of the war campaign in 2009. My decision to work in the east was enabled by the government's newfound control of the region, but was largely motivated by a desire to learn about a region referred to as the "crucible of conflict" (McGilvray 2008) in Sri Lanka, still recovering and reconstructing after years of trauma.

As I will further detail in chapter 1, I was interested in this region in particular because of its complexly layered social and political histories of violence, migration, displacement, and resettlement. Being predominantly Tamil and Muslim, the east has a history of being marked by border zones, front lines and no-man's-lands from war (de Alwis 2004; Hyndman and de Alwis 2004). My research was concerned with how new spaces and complexities engendered by the tsunami and national disaster management mapped onto the livelihoods of coastal communities already affected by years of war (see map I.4). The east coast was the area hardest hit by the tsunami, as the region closest to the epicenter of the tsunami-generating earthquake and also the most densely populated place on the island. For many years, and especially during the 1990s, the Eastern Province was much fought-over territory. Along with the Northern Province, the Eastern—where the Tamil language is dominant—is considered part of the LTTE's problematically imagined monoethnic homeland and during the war was controlled by them, until 2007, when government forces "liberated" the east.

I could not have done my research in the east without the company and knowledge of my two research assistants, Usha and Nitharshini, who also helped to translate from Tamil the complex and moving stories shared in this book. Smart, ambitious women from the east, with great intuition, they also offered their opinions on my fieldwork ideas, on where to go and with whom to talk. After hearing many sad stories in one day, Usha told me that even though she had heard so many stories, the "route of the sadness" was always different. I have tried to follow those routes with care. I maintain contact with her and with other friends who are able to provide me with general updates, community feelings, and other local insights that are not always conveyed in national news outlets. When I left in late September 2009, Sri Lanka's post-conflict situation was still tense. The state of emergency had just been extended by Parliament, Tamil civilians were still interned in camps in the north, and the government had increased the military budget by 20 percent and increased the size of the military by 50 percent, all in the name of protecting its fragile "security."

Over the course of my field research, I worked with disaster management practitioners, community leaders, and communities of people displaced by both the tsunami and the war. Given such complex histories of destruction and adversity, the following chapters paint an intimate portrait of how life persists under conditions of perpetual threat—from either another natural disaster or another outbreak of state-sanctioned and war-related violence. Many of the people I came to know shared with me their experiences during some of the worst moments of fighting between government and LTTE forces and their current experiences and interactions with police and military personnel. I also met regularly with local and district-level disaster management officers and coordinators, participating in community disaster preparedness workshops and evacuation drills. Their dedication and care for their communities was admirable. I spent time in several coastal villages across the Eastern Province, mostly near Kalmunai, the largest city on the east coast and one of the few Muslim-majority municipalities in Sri Lanka. I interviewed district-level planners, the mayor, and even the special branch of the police, the Special Task Force (STF), whose work continued because Sri Lanka's state of emergency was still in effect even after the government's declaration of peace. I also interviewed local and international humanitarian aid and development workers conducting reconstruction projects in the region.

At the national level, I made regular visits to the National Disaster Management Centre in the capital city, observing disaster protocols and learning about the technical coordination of the national disaster warning system. I met with meteorologists, disaster management officials, national coordinators, and systems technicians, whose efforts and genuine concern about disasters and disaster preparedness was evident. I also attended national rallies in support of the government's war efforts and victory. My fieldwork gave me the opportunity to examine different forms, fantasies, and scales of nationalist projects unfolding during this significant moment in Sri Lankan history. I also acknowledge that despite the complex security situations across the island, my positionality as a perceived "white" (*vellai* in Tamil, *sudhu* in Sinhala) researcher from the United States allowed me to move through spaces without garnering suspicion (experiences of immobility in the east I discuss in more detail in chapter 3).[21]

DISASTER AS/AND NATIONAL FANTASY

> By "fantasy" I mean to designate how national culture becomes local—through the images, narratives, monuments, and sites that circulate through personal/collective consciousness.
>
> —Lauren Berlant, *The Anatomy of National Fantasy*

The day following President Mahinda Rajapaksa's official announcement of war victory was declared a national holiday. When the war ended, I was in the east. Streets were quiet, and the mood in town was decidedly somber. The television, however, displayed another story. In other parts of the island, the Sinhala south and the capital, Colombo, residents had erupted in celebration. On the streets, cauldrons overflowed with *kiribath* (milk rice), a food traditionally prepared for auspicious occasions and for the Sinhalese New Year. Throngs of people crowded streets, their boisterous shouting enlivened by the banging of drums and popping of firecrackers. On-screen images were saturated with the golden and crimson hues of the Sri Lankan flag, the Sinhalese Lion proudly on display. These cultural emblems of the nation became more palpable with the multiplying proliferation of flags displayed on shops, homes, and three-wheelers as I made my way from the Tamil and Muslim east, through the predominantly Sinhalese south, and further westward to the capital, Colombo. Traffic was slow entering the suburbs of the capital while weaving through the rambunctious and clamorous crowds. I had entered the circulating national fantasy that I had previously watched on TV.

National fantasies are not illusory constructions and cannot be cast aside merely as ideologies. As Berlant (1994) notes, they become localized and embedded in collective consciousness; they are forms of reality with real, material effects (Aretxaga 2003; Navaro-Yashin 2002; see also de Mel 2007). The technologies of disaster nationalism employed by the Sri Lankan state are legitimized through the discursive, imagined, and objective components of national fantasy: the securitized, united, and liberated nation; the impending future of disasters and catastrophe; the rallying around a common enemy (Thiranagama 2022).[22] These fantasies of nationalism erase histories of disaster such that by 2010, Sri Lanka's "pristine coastlines" could be the *New York Times*' number one place to visit, victory memorials could be erected and LTTE cemeteries in the war-torn north could be razed, and President Rajapaksa himself could disavow the violent end of the civil war.[23] In an interview with *Time* magazine entitled "The Man Who Tamed the Tamil Tigers" (Thottam 2009), Mahinda Rajapaksa was asked if a Truth and Reconciliation Commission would be implemented in the aftermath of the war, to which he responded, "I don't want to dig into this past and open up this wound."

The pristine, romanticized fantasy production (Tadiar 2004) of a peaceful Sri Lanka and the desire to quickly move forward while burying the past were mobilized and fed into this new imagination of the nation, sustaining the power of President Rajapaksa's governmental and military regime. Indeed, this imagination gained traction, and was needed, in part because Sri Lanka did not have a "normal" past to which to return. Sri Lanka's "normal" was a seemingly interminable war, around which lives, politics, and economies had been organized (Rajasingham-Senanayake 2001; Ruwanpura 2006; Somasundaram 1998, 2013; Winslow and Woost 2004). The future nation, in turn, could be marshaled by the Sri Lankan government as both a nation of endless possibility and growth and a vulnerable nation in need of protection, its fragile order always at risk of being punctured by natural or human-made catastrophe. With an eye cast toward a future of foreign investment and tourism, Rajapaksa's regime presumed that this new Sri Lanka would be poised to cash out on the peace dividend.

Fantasies are also crucial to developing national affects and imaginations of disaster. In the United States, as Joe Masco (2008) recounts, Cold War civil defense strategies created a new social contract based on a national contemplation of ruins. Building the bomb and communicating its power through epic images of nuclear explosions and annihilation introduced a collective national consciousness of destruction and death. As he further explains, by the 1950s, witnessing nuclear annihilation had become a "formidable public ritual"—a core act of governance, technoscientific practice, and democratic participation. Highlighting the legacies of these Cold War international and national strategies, Masco argues that these "willful fabulations" and "official fantasies" were politically useful, generating an affect of "productive fear" and an endangered nation through the imagination of it in ruins (Masco 2008). In Sri Lanka, however, a collective consciousness of destruction did not necessarily need to be cultivated because for many, the unimaginable, the impossible had already happened. Rather, the ruins of the past would have to give way to a new vision of the nation: a Sri Lankan nation prepared for disaster and freed, finally, from the scourge of terror.

Sri Lanka's fragile peace, then, necessitated the securitization and protection of territory and infrastructure. In Sri Lanka, certainly, war-related infrastructures were in place, yet other everyday infrastructures had long been neglected or affected by the war. The task of "Building Back Better" and projects of disaster nationalism were not necessarily aimed at protecting existing infrastructures, but rather at building new ones. Buffer and border zones, national disaster early warning systems, military checkpoints—these technologies of securitization and modes of preparedness materialized the construction of the newly liberated yet always endangered nation, inviting state intervention.[24] The nation thus becomes the object and purpose of struggle, management, and control. The perception of imminent disaster, whether tsunamis, earthquakes, civil unrest, or terrorist attacks, becomes a technology of power that continually conjures national fantasies of securitization (Heath-Kelly

2018). National fantasy "thus entails a certain sense of security, fixity and fullness by rationalizing the past, justifying the present and prescribing for the future" (Mandelbaum 2019, 22).

DISASTER CHRONOPOLITICS

> It is important to remember that historical events are never really punctual—despite the appearance of this one and the abruptness of its violence—but extend into a before and an after of historical time that only gradually unfold, to disclose the full dimensions of the historicity of the event.
>
> —Fredric Jameson, "The Dialectics of Disaster"

> Disasters aren't events that float freely in history, unmoored from politics: they are processes, playing out in uneven temporalities, and always with deep histories.
>
> —Scott Knowles, "Slow Disaster in the Anthropocene"

The "historical event" Fredric Jameson refers to in the aforementioned quote is the September 11 attacks on the World Trade Center. The publicly shared sentiment of national devastation in the United States proffered through media spectacle, he and Knowles remind us, should not be presumed or viewed as self-explanatory, unmoored from history and politics. Time is a technique of power (Bear 2016; Gutkowski 2018; Zee 2017); history and futurity can be manipulated by the state in moments of crisis to legitimize institutions of law and politics (Greenhouse 1996; see also Alonso 1994).[25]

Thinking with the vast scholarship of nationalism in Sri Lanka and beyond, alongside the growing body of work on disaster, risk, and insecurity, the preemptive temporality of disaster and risk management converges with the temporality of nationalism.[26] Inspired by Walter Benjamin, Benedict Anderson characterized collective national consciousness as the time of capital, or homogeneous, empty time (Anderson 1991); he proposed that the nation was made possible by imagining other citizens, near and far, as living and existing in the same national space and moving together through a shared sense of linear time and a shared sense of history. In Sri Lanka, the homogeneous time of disaster nationalism is more than a collective consciousness or shared sense of linear time. Nationalism is not only upheld by traditional, timeless narratives of the past but also maintained through the insistence of a future horizon of disaster (Kim 2016; Masco 2014). In Sri Lanka the homogeneous time of disaster nationalism hinges upon and imposes a future-oriented time of securitization, preemption, and preparedness (Anderson 2010).

Yet time, as Sarah Sharma reminds us, "is lived at the intersection of a range of social differences" (2014, 138). As such, I resist folding experiences of disaster into the teleological history and narrative of disaster nationalism. As John Kelly (1998) points out, Walter Benjamin too refused this linear image of history. The lived realities and experiences of Sri Lankans I came to know complicate the chrononormative (Freeman 2010), homogeneous composition of the imagined community of the nation, illustrating that it is, rather, fragmented and heterotopic (Chatterjee 2004, 2005; Duara 1996; Foucault 1984; Wickramasinghe 2014)—an unattainable political project that has incommensurability and exclusion at its core (Hansen and Stepputat 2006; Jeganathan and Ismail 1995; Lomnitz 2001; Sur 2020). As Beth Povinelli writes, "we find that although all people may belong to nationalism, not all people occupy the same tense of nationalism" (2011, 38): the nation has never been homogeneous (Spencer 2003; Tsing 2011; see also Latour 1993). State desires for national domination, while powerful and comprehensive, are never absolute or complete.

Each chapter of this book traces how national security fantasies play out, and the emergent forms of life they enact and engender (Fischer 2003; see also Thiranagama 2022). The chapters that follow examine these generative forces of state power after and amid disasters, as well as the ways in which people *endure* the experience of disasters and the effects of disaster nationalism. I look for the articulation of these experiences and negotiations of power in maps, in memories, in tsunami warning towers, in military checkpoints, in broken foundations of old homes, in hot tin shacks in which people await their new tsunami homes (Navaro-Yashin 2012). As Kathleen Stewart (2007) offers, the mundane and the "ordinary" are the very building blocks of life. What is it like to live with both the fetishization of catastrophe and the everyday negotiation of catastrophe that renders it banal? This ethnography holds both the spectacular and the mundane together, for the spectacular in Sri Lanka is also the mundane. The disaster is multiple (Mol 2002).

These questions point to the tempos and rhythms of living with disastrous futures, the ways that the future catastrophes of the *now* are tempered by the possibilities of survival. In Sri Lanka, enduring disasters is a condition of possibility, of possible futures and pasts that might differ from the national contemplation of catastrophe (Das 2007; Honig 2009; Puar 2007).[27] Kathleen Stewart writes: "Things do not simply fall into ruin or dissipate.... [They] fashion themselves into powerful effects that remember things in such a way that 'history' digs itself into the present and people cain't [sic] help but recall it" (Stewart 1996, 111). The past offers a condition of possibility, of speaking back to history and the present and keeping them open (Yoneyama 1999), of unsettling that which seems settled (Fortun 2001). The tsunami and the war are certainly different, as are their legacies, but their histories run into each other, layer upon each other—they are both evoked through the reconstruction of the nation.[28] A new post-disaster peace was not a *terra nullius* (Klein 2005) upon which the damages of the past could be forgotten.

Security, violence, militarism—these are familiar terms for many scholars of Sri Lanka and beyond, who have shown how violence characterizes everyday aspects of life and who importantly point to the forms of power embedded in the social and political structures that create the conditions for inequality, injury, and suffering.[29] Violence manifests not only in spectacular or episodic moments but also in subtler, slower, and more ordinary ways. Too much of a focus on discrete events or crises that invite political intervention draw attention away from the fact that crises are constitutive of contemporary social life.[30] In tsunami- and war-affected Sri Lanka, moving beyond a framework of risk and vulnerability, I trace how disaster is also constitutive of life and politics, and how disaster nationalism perpetuates existing forms and conditions of precarity and insecurity in a society where violence and militarization have already long characterized life (Choi 2021; de Mel 2007; Hewamanne 2013). It is the very ordinariness of insecurity and violence that is disastrous (Thiranagama 2011). This book details how Sri Lankans have persevered through the tsunami and outbreaks of war-related violence, and continue to negotiate the possibilities of natural disaster or state-sanctioned violence. Here, disaster as event or nonevent is spectacular *and* banal, punctuated *and* chronic (Das 2007; Massumi 2005). The persistence of insecurity in Sri Lanka shows that the beginnings and endings of disasters cannot be easily demarcated (Samuels 2019).

Each chapter of the book juxtaposes the tsunami and the war, invoking a modality of time. They can be read in order or out of order, suggesting that these modes of time exist contemporaneously, in concert with, against, and alongside the temporal horizon of impending disaster, of the impending failures of an authoritarian state and the unrealizable fantasies of nationalism that give structure to its power (Holbraad and Pedersen 2013). In these overdetermined spaces of disaster—these spaces of militarization, death, economic insecurity, tsunamis, and cyclones—what of Sri Lankan lives and experiences? Social experiences of time are multiple and uneven (Bear 2016), evincing the complex and undisciplined chronopolitics of life for those impacted by tsunami and war in eastern Sri Lanka. Recognizing this chronopolitics is not just a gesture toward an apolitical pluralism, but an effort to reconcile the vestiges of past violences with the traumas of the present and imaginations of the future. That is, if the nation has never been homogeneous, these chapters make clear that the political ordering of disaster nationalism that seeks to colonize the future and the past is always a fraught and fricative project (Tsing 2005). The nation-state as an inherently unstable project works to both undermine and justify modes of governance and securitization in Sri Lanka.

Chapter 1, "Emergence," provides more context regarding Sri Lanka's fraught postindependence politics, with a particular focus on the east of Sri Lanka, where much of my fieldwork was conducted. The chapter's inspiration came from an attempt to research the 1978 cyclone that devastated the eastern coast of Sri Lanka, the country's biggest natural disaster before the 2004 tsunami. The year 1978 in particular was a watershed for Sri Lanka: a new constitution; a new president; the state-authorized liberalization of the economy; and the development of an anti-terrorism bill. While much has been written about this particular time period's political and economic shifts, I have never found the cyclone referenced, much less figured meaningfully into, these accounts. The chapter does not attempt to insert the cyclone into these histories, but rather explores its absence in order to trace the *emergence* of legacies of state governance leading up to the Sri Lankan government's disaster management practices after the tsunami. Given Sri Lanka's history of war, the militaristic turn after the tsunami is perhaps no surprise. The chapter uses "disaster" as a lens to highlight how disaster nationalism is a new dimension of national governance and state power and also is contiguous with a longer history of institutionalized exclusion and Sinhala Buddhist majoritarian politics and nationalism. Finally, drawing a longer history of political "disaster" in Sri Lanka underscores the recursiveness of violent nationalist politics on the island, illustrating the difficulties in locating the beginnings and ends of disaster.

The following chapter, "Anticipation," traces how the notion of future disaster is negotiated. I show how, amid momentous social and political changes in Sri Lanka, many people have experienced a seemingly palpable lack of change. Despite a cessation of war, I found that an anticipation of violence persists. I depict how it is to live with that specter of violence, in addition to the possibility that life can also be disrupted, taken, broken by a natural disaster. These anticipations are part and parcel of everyday life in Sri Lanka. I weave together various states of anticipation—from a mode of technocratic governance, to the everyday emotional and sensory infrastructures people employ to face daily life, to geology and to astrology. By juxtaposing these "anticipatory states," the chapter illustrates how forms of state power are enacted and articulated, and how the limits of these powers are exhibited by the very practices of the Sri Lankan state.

Amid this sense of ongoing or ever-present disaster, the next chapter, "Endurance," highlights the texture and rhythms of life on the east coast as they unfold in the everyday, in places where death and destruction have been so swift and sweeping, in places still highly militarized after the tsunami and supposed liberation from terrorism. In this chapter, I develop the notion of "slow life" to draw attention to the ways in which life—not just in the biological sense—persists in contexts of insecurity and social and political anxiety. The chapter contains two main sections that follow these enduring forms of slow life. First, I detail the experiences of those living in a temporary post-tsunami housing scheme and awaiting their newly built tsunami homes, nearly five years after the tsunami. In the second section, I shift ethnographic form to present a series of text messages distributed by the Sri Lankan government, coordinated with the Ministry of Defence, to provide updates on their war efforts. I juxtapose these messages with excerpts from daily conversations taken directly from my field notes corresponding to the dates these messages were sent. This section and these juxtapositions more performatively illustrate the chronopolitics of disaster nationalism: enduring forms of (slow) life in a disaster-obsessed state. The chapter moves through these modes of endurance in the everyday: waiting for new homes; waiting at a checkpoint; waiting for the next tsunami; passing time; making conversations—alongside, within, and sometimes against the national narrative of catastrophe and erasure of the past.

Chapter 4, "Reiteration," explores iterative productions of the Sri Lankan nation through the visual culture of the tsunami and the war. I highlight how national politics are continually made and unmade through the production, use, reuse, and distribution of images of the tsunami and the war.

After the tsunami, images and maps were given much credence and heralded for their value in representing damage and destruction. By contrast, maps of the conflict became highly fraught, illustrating different political interests. Because the Sri Lankan government forbade the presence of independent journalists and international organizations into the war-torn north, rumors of the "ground truth" swirled in both local and international media. During this time, the unknown became a productive political moment. The chapter follows images' unruly and unstable qualities as they were attached to various political modes, motives, and contradictions of truth-making in the war by governmental and nongovernmental regimes. In doing so, I show the ways in which disasters live on through the contestations and politics surrounding images and how, after May 2009 in Sri Lanka, peace became the continuation of war by other means (Foucault 2003).

Brush and sand cover upturned wells and broken foundations, as new houses are built on top of liberated landscapes where previously mines were buried, where blood was shed, where lives were swept away. I saw the wreckage of the past all around me as President Rajapaksa claimed that he "did not want to dig up the past." This selective amnesia upholds the victor's vision of a united Sri Lankan nation and attempts to foreclose the narratives and complexities that have constituted it.

The chapters that follow keep these narratives open, for, as Saidiya Hartman reminds us: "History is an injury that has yet to cease happening" (2002, 771). Working through other modalities of time disrupts the circular, future-oriented, and linear temporality proposed by disaster risk management frameworks—Mitigation → Preparedness → Response → Rehabilitation → Mitigation—drawing attention to and challenging recursive and enduring disaster teleologies of nationalism in Sri Lanka.[31]

Come. Let us sift through the rubble.[32]

You must understand: There is no single day on which a war begins.

V. V. Ganeshananthan, *Brotherless Night*, 2023

Fighter jets at the National Victory Day Parade, June 3, 2009. Photo by author.

Starting out with emergence as a question is also valuable because, in addition to asking what is new on the horizon, it suggests that contemporary practices are unfinished, ongoing, continuously maintained.

Gary Lee Downey and Joseph Dumit, "Locating and Intervening"

1 Emergence

ONE NATION, UNDER DISASTER

To celebrate the end of the war, the government organized a Victory Day Parade (the name of the day has since been changed to War Heroes Day) and National Tribute to Security Forces on Galle Face Green in Colombo. The parade was an opportunity to display the Sri Lankan military might that had decisively defeated the Tigers. On June 3, 2009, Sri Lankan Army commandos marched in step; soldiers carried Sri Lankan flags, brandishing waves of mustard yellow and crimson; army tanks slowly rumbled in line; large military trucks pulled heavy artillery used for shelling in the war; as pictured in the image preceding this chapter, the Sri Lankan Air Force's jets swooped through the almost-cloudless blue sky. Before the military fete began, motorcades of cars with black-tinted windows transported parliamentary members and high-profile Sri Lankans close to the president to the site, but Mahinda Rajapaksa, flanked by guards and high-ranking officials, rode onto the Green in a topless jeep, waving at the crowds. I had left the east to experience how different the response to the end of the war was in the south. It startled me to be this close, as I had only ever seen the president's image on television and in newspapers. To be part of the crowd of boisterous parade-goers, my friend and I had gone through several security checkpoints (including complete body pat downs) to earn our "security checked" stickers. Hawkers sold tiny Sri Lankan flags and even posters and stickers of Rajapaksa's face, one of which I purchased. Rajapaksa (2009) also addressed the nation, first speaking in Tamil:

> Friends,
>
> This is the Motherland of us all.
>
> We should live in this country as the children of one mother, as brothers and sisters.
>
> There can be no differences here. The war fought against the LTTE was not a war fought against the Tamil people.
>
> Our Heroic Troops sacrificed their lives to save the innocent Tamil people from the clutches of the LTTE. We cannot forget the great service rendered by them.
>
> The victory we gained defeating the LTTE, is a victory for our entire land.
>
> It is great victory you obtained.
>
> It is a victory for all who live in our country.
>
> The war against the terrorists is now over.
>
> It is now the time to win over the hearts of the Tamil people. The Tamil speaking people should be protected. They should be able to live without fear and mistrust.
>
> That is today the responsibility of us all!

The rest of the speech went on to praise Sri Lanka's "heroic troops," mentioning the country's "tradition" of facing "invaders":

> This is a country with a people having a tradition of facing the most ruthless invaders through thousands of years, defeating them, and saving our Motherland. Our country is home to a people with a history of bravely facing up to invaders from the time of King Dutugemunu to the last king of the Sinhala Kingdom, Sri Wickrama Rajasinghe, and fearless patriots such as Keppetipola and Puran Appu.
>
> The lessons we learnt from those great battles of the past are ingrained in our flesh, blood and bones. Our brave soldiers were not provoked, but waited with discipline till the proper time for action came.
>
> We can now see how much determination is born from the pain one suffers.... What was considered the most ruthless and most powerful terrorist army for thirty years was shattered and destroyed in less than three years.

> Terrorists are no more invincible. It is only the valiant troops and our Motherland that are invincible.

These words echoed a speech Rajakapaksa delivered to Parliament in May, immediately following the death of LTTE leader Prabhakaran and the end of the war, in which he also detailed Sri Lanka's "long history" of invasions:

> We are a country with a long history where we saw the reign of 182 kings who ruled with pride and honour that extended more than 2,500 years. This is a country where kings such as Dutugemunu, Valagamba, Dhatusena and Vijayabahu defeated enemy invasions and ensured our freedom.
>
> As much as Mother Lanka fought against invaders such as Datiya, Pitiya, Palayamara, Siva and Elara in the past, we have the experience of having fought the Portuguese, Dutch and British who established empires in the world. As much as the great kings such as Mayadunne, Rajasingha I and Vimaladharmasuriya, it is necessary to also recall the great heroes such as Keppettipola and Puran Appu who fought with such valour against imperialism.
>
> In looking at this unconquerable history there is a common factor we can see. It is the inability of any external enemy to subdue this country as long as those to whom this is the motherland stand united.
>
> That is the truth.

In this call for, or emergence of, a "new patriotism" (Wickramasinghe 2009), there are no longer any minorities, simply "those who love this country" and "those that have no love" for the land (Rajapaksa in Wickramasinghe 2009, 1046). Yet simultaneously, the speeches invoke an ethnicized historical and mythic consciousness that valorizes an "unconquerable" Sinhalese Buddhist nation, imagined as existing since time immemorial. Rajapaksa seamlessly conflates the contemporary modern form of the nation-state with ancient, premodern forms of social and political power and organization in Sri Lanka. What's more, the "great" kings that President Rajapaksa notes as heroes and defenders of Sri Lankan freedom, Dutugemunu, Valagamba, Dhatusena, and Vijayabahu, are Sinhala Kings who fended off Tamil-Dravidian invaders Siva, Elara, Pitiya, and Datiya from Southern India. Tamil "invaders" are also lumped in with Portuguese, Dutch, and British colonizers as external enemies that have attempted to "subdue" and rule the motherland of Lanka. Rajapaksa's speech thus fixes the Sri Lankan national imaginary in space (motherland) and time (history), in an attempt to conjure a "pure" Sinhala Buddhist national time-space (Jazeel 2013; Tennekoon 1983), perpetually threatened by enemy invasions.

Rajapaksa was not the first president to publicly put forth a notion of an unbroken 2,500-year-old Sinhala Buddhist history and act upon it.[1] In his speech, President Rajapaksa reidentifies and marks who is an outsider, or an external threat to the security and unity of the glorious (Sinhala) country. Yet, if these statements are intended to unite the Sri Lankan motherland against "terrorism" or the LTTE—threats to national security—they elide the fact that the LTTE was an enemy that emerged from *within* Sri Lanka, originating from the nation-state's own discriminatory practices *against* its own people. As this chapter will detail, state-sponsored nationalist projects have harkened materially and symbolically to a Sinhala past while also signifying aspirations for the future, in part through marking a national "other" or an assertion of the nation's "constitutive outside" (Valluvan 2019, 36). This chapter considers this mythic nationalist narration of Sri Lankan history and its significance to the management of disaster. That is, it historicizes disaster nationalism.

The emergence of disaster nationalism in Sri Lanka after the tsunami, in light of the war, may not be surprising. Given decades of insecurity, war-related violence, state repression, and a nearly continuous state of emergency since 1971, in addition to the primacy of the "Global War on Terror," the Rajapaksa administration's militarization of disaster and its aggressive approach to eliminating the LTTE was perhaps even expected. Expected, perhaps, but not inevitable. What precedents—social, political, institutional—were in place such that by the end of the war, under the mechanisms of disaster nationalism, tsunamis *and* terrorism could both be perceived and governed as disasters and threats to national security? How did disaster management also become a mode of national securitization? The temporal theme of the chapter, "emergence," is a processual approach, a way to attend to the emergence of Sinhala nationalist imaginaries and the emergence of threats to mythic imaginations and fantasies of nationalist purity that Rajapaksa evokes above in his speeches. The chapter uses "disaster" as a lens to trace a new dimension of national governance and state power through disaster nationalism while *also* highlighting disaster nationalism's contiguity with a longer history of institutionalized exclusion and Sinhala Buddhist majoritarian politics and nationalism.

To understand this new dimension of state power and nationalist hegemony, the following section explores a watershed moment in Sri Lanka's recent history: 1978. This was a remarkable year in Sri Lankan politics. Its beginning had ushered in a new president, J. R. Jayawardena, who was empowered by a new constitution and a new political majority, the UNP, in Parliament. Other significant events in 1978 included the opening up of Sri Lanka's economy; the creation of the Proscribing of the Liberation Tigers of Tamil Eelam and Similar Organizations Act No. 16 in May; and the development of the Prevention of Terrorism Act (which would be passed as a temporary measure in 1979 and made into permanent legislation in 1982). While much has been written about this political, constitutional, and economic moment—including the social tensions culminating in the beginning of the war in 1983—one event is generally absent from these narratives: the 1978 cyclone which devastated the eastern coast and was the worst natural disaster in Sri Lanka before the tsunami. I do not attempt to insert the cyclone into these analyses; rather I explore what its absence can tell us about how the tsunami merged with Sri Lankan social and political life. That is, investigating the cyclone opens up a broader political history of disaster that is at once specific to the east and also consequential to national politics.

The next section provides further institutional and nationalist history, highlighting the emergence of discriminatory notions of "aliens" and "unassimilated" and "undomiciled" others in the postindependence Sri Lankan nationalist imagination, as part of state-sponsored efforts contributing to the emergence and dominance of Sinhala Buddhist nationalism. This history of disaster (nationalism) shows what the war, militarization, and nationalism makes available for the state over time and how it structures the lives and experiences of Sri Lankans. This history helps the broader intentions of the book by tracing the tensions between, on the one hand, ideas of disaster and the nation produced through the mechanisms and processes of disaster nationalism and, on the other, Sri Lankan experiences and life, which insistently disrupt these fantasies of disaster, the nation, and nationalism.

"Emergence," as referenced by Downey and Dumit (1997), is a focus on the processual. As such, this chapter does not consider Sri Lanka's historical trajectory and ethno-nationalist violence as inevitable or determined, but rather focuses on the continual efforts—of state power and maneuvering—required to suffuse nationalist hegemony across so many dimensions of life (Hansen and Stepputat 2006; Jessop 2005; Rampton 2011; Williams

1977). As Stuart Hall writes, "No victories are permanent or final" (2011, 26). While emergence connotes the catastrophe that was nearly forty years of a state of emergency in Sri Lanka, recognized as a "state of permanent crisis" (Welikala 2008; see also Coomaraswamy and de los Reyes 2004; Hewage 2014; Udagama 2015), it also refers to other forms of emergency governance beyond or not exclusively within a paradigm of exceptionality. The radical ontological contingency of life demands new forms of governance, requiring constant adaptive emergence of state power and rationality (Dillon and Lobo-Guerrero 2009; see also Adey, Anderson, and Graham 2015). This is not a comprehensive history. This is not an argument outlining the root causes of the war. "Emergence" questions the homogenizing forces of disaster nationalism. That is, it questions the nature and naturalization of war, disaster, and threats to the nation—powerful, violent structural forces across the island and especially in the east.

THE 1978 CYCLONE AND THE EAST

In late November 1978, a cyclone battered the eastern coast of Sri Lanka. As an elderly Tamil man recounted to me, in the morning after the storm had passed and the winds had calmed, he looked inland. He said he had a view into the country as far as the eye could see. Other cyclone survivors also recalled this stark transformation after a dark night of howling winds and torrential rain: the cyclone had flattened everything in its path such that nothing impeded the view of the land. Torrential rains and wind gusts of up to 160 miles per hour destroyed over 100,000 homes and damaged another 85,000. The storm killed 740 people and injured 5,000. Public buildings including hospitals and schools were damaged. Seawater inundated land 5–10 kilometers from the coast, destroying agricultural land—the eastern coconut plantations never recovered—and damaging other infrastructure (Office of Foreign Disaster Assistance 1979). This was Sri Lanka's worst natural disaster before the 2004 tsunami. Certainly in scale the cyclone was smaller, but its devastation at the time was unprecedented.[2]

Selvi told me her memories of the cyclone: she was very young and lived with her family in what is now a predominantly Muslim area of Kalmunai. As the day darkened early, the wind swirled. They took shelter in a neighbor's home. Unlike most others that were made with *cadjan* (interwoven palm leaves), their home was built with concrete and therefore better able to withstand the cyclone's heavy rain and winds. The home belonged to a Muslim. I was surprised that Selvi, a Tamil woman, had been living in this Muslim neighborhood. She explained to me that "in those days" the neighborhoods were mixed, though she commented, "today, we can't even imagine" these kinds of living arrangements. Asking others about their cyclone experiences, I heard similar accounts. People took refuge where they could, and neighbors opened their homes to neighbors in need of shelter. Neighbors also helped each other clean up debris and rebuild homes. As was relayed to me, several families felt that assistance, whether from the government or from international organizations, was slow. And that assistance, when it did arrive, mostly came in the form of food, not housing.

Selvi's comment struck me—she understood that the conflict had changed life and communal relations in the east, so much so that past ways of neighborhood arrangements and living conditions would now be "unimaginable." Later, my Muslim friend Raheem, recounting his experiences of the cyclone, told me that those days, he was one of three Muslim boys that went to school in a primarily Tamil area of Kalmunai, but that they all still played together. He said Muslim families would give refuge to Tamil neighbors running from the STF or the Sri Lankan Army. He said, "It is only now that whenever there is a problem, we think and join up along our religious or ethnic lines." Selvi and Raheem understood that the war had, over time, changed the landscape in which they lived and even their imaginations.

Segregation between Muslims and Tamils within towns and divisions along the east coast occurred gradually during the war (Fonseka and Raheem 2010; McGilvray 2001, 2008; Walker 2013). Resettlement policies and practices after the tsunami also created considerable tension along ethnic, caste, and social class lines, in a context where finding alternative sites for communities was hampered by land scarcity and security concerns, particularly for tsunami-impacted Muslims (Hasbullah and Korf 2013; Silva and Hasbullah 2019). These social tensions, as I further describe below, are part of a longer history of social tensions in the east, cultivated not just through the conflict, but over decades of state-sponsored land colonization.

The Eastern Province spans across Trincomalee District, Batticaloa District, and Ampara District. In the eastern coastal region of Sri Lanka, the A-4 road runs along the coastline. Headed north, on one's left, rice-paddy fields stretch seemingly forever into a horizon of tall grass. The Eastern Province is the agricultural heartland of Sri Lanka, included in what is called the "Dry

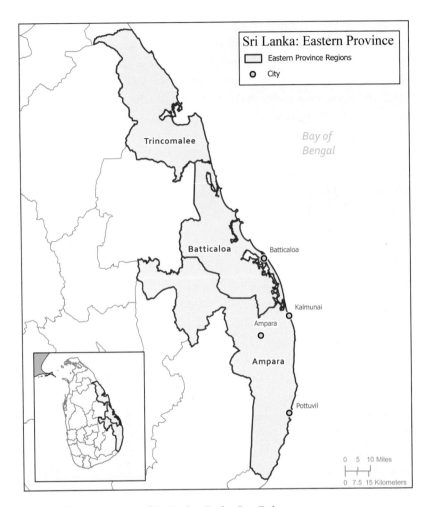

MAP I.I. Eastern Province of Sri Lanka. Credit: Sara Dale.

Zone." On one's right, homes, temples, police stations, and post offices dot the landscape. Beyond buildings is the deep azure of the Indian Ocean, lapping at the embittered shoreline of what has been referred to as Sri Lanka's "crucible of conflict" (McGilvray 2008; see also Spencer et al. 2014). The east, once included in the LTTE's geographically imagined ethnic homeland, was the stage for many war atrocities committed by both the government and LTTE. Away from the hustle and bustle of main town areas, where horns, exhaust, and noisy motors bombard the senses, the interior roads nestled between different villages are comparably quiet. The setting for my research was the largest city in the Eastern Province, Kalmunai, in Ampara District (see map 1.1 above).

The Eastern Province is the most diverse, ethnically and religiously, in Sri Lanka. Its diversity, while problematizing both Sinhala Buddhist and Tamil homogeneous nationalist imaginations, has also made it a complex and fraught territory historically and politically amid the ethnically polarizing war. Nationalist LTTE proclamations of the northern and eastern parts of the island as solely for Tamils completely ignore the long history of both Tamil and Muslim settlement and sociality in those regions. Moreover, the north, or Jaffna, has always figured as central to the imagination of Tamil identity and to its imagined homeland. This has been a point of contention for eastern Tamils, who feel that they are discriminated against and express distrust at the thought of a separatist state controlled by the LTTE (McGilvray 2008; Thiranagama 2011; Walker 2013; Whitaker 1997). Muslims make up the eastern coast in equal if not larger numbers than Tamils. The polarization of the conflict has put them in the difficult, if ineluctable position of having to shift alliances in the battle between the dominant Sinhalese majority and the Tamils. As Ismail, Rameez, and Fazil (2005) point out, as the conflict escalated, Muslim communities in the north and the east were exposed to terrorism by both Tamil militants and the Sri Lankan state (see also Haniffa 2008).[3]

Dennis McGilvray's (2008) extensive work in the east documents the dynamic inter-ethnicity and religiosity of the region, especially on the coast. A shared heritage between Muslims and Tamils in the east is exemplified in marriage and caste practices, in addition to a long history of social exchange between Tamils and Sinhalas, offering a perspective on social diversity in Sri Lanka that contrasts with the "diversity" created through postindependence political gerrymandering and state-sponsored colonization. For example, the Gal Oya Irrigation Scheme in 1949 undertaken by Prime Minister D. S. Senanayake introduced landless peasants—most of whom were Sinhalese—into newly developed and irrigated land near the dammed Gal Oya river. While on the surface these development schemes aimed to address poverty, agricultural production, and the generation of electricity, they were also used to galvanize Sinhala nationalist sentiments (Moore 1985), such that the success of colonization of the Dry Zone could be seen through changing ethnic demographics in the region and success was measured in social rather than financial terms (Rambukwella 2018). Around this time, civil unrest was unfolding in Colombo related to the Parliament's passing of the 1956 Sinhala Only Act; mob riots broke out against Tamils staging a sit-in on Parliament steps.[4] When rumors of the riots spread to the east, agitated Sinhalese mobs in Gal Oya began roaming the streets looking for Tamils. It is estimated that nearly one hundred Tamils were killed during this time (Tambiah 1997; see also Manor 1984). In the 1960s, the Eastern Province was officially created, carved out of a larger Tamil-dominated region which included Batticaloa in the northeast. By adding Ampara town further south, the predominantly Sinhalese inland capital and the new administrative and trade hub of the Eastern Province, including Gal Oya, the government broke up any ethnic majority in the east and prevented the dominance of a monolithic voting block. Such diversity, whether enforced by the state or through local norms and traditions, has resulted in a "variety of geographies of antagonisms... across different political registers of community and enmity (inter-ethnic and intra-ethnic, inter-religious and intra-religious), which have trickled down to smaller territorial scales: settlements, enclaves, divisional secretariat divisions" (Hasbullah and Korf 2013, 34).

The late 1970s, when the cyclone devastated the east, also marked a tumultuous, eventful period in Sri Lanka. In 1977, immediately after the election of J. R. Jayawardena—who also claimed that he was Sri Lanka's 193rd head of state and king of Sri Lanka (see Krishna 1996)—anti-Tamil riots broke out around the island, and particularly in the east, in reaction to rumors that the newly formed political party the Tamil United Liberation Front (TULF) planned to realize a separate Tamil state. Meanwhile in the north, a violent and militant faction of young Tamil men in Jaffna—who would later become the LTTE—were robbing banks and killing policemen. In 1978, a new constitution was introduced, establishing an executive presidency for the first time. This gave Jayawardena and future presidents the power to declare states of emergency, order armed forces to maintain law and order, and absorb the role of minister of defence and issue detention orders under the Prevention of Terrorism Act, also developed in 1978 and passed in 1979 (International Commission of Jurists 2012).

With the election of Sri Lanka's first president, J. R. Jayawardena, in 1977, "the level of violence and the pace of colonization in the Dry Zone between the Sinhalese and Tamil majority areas increased" (Peebles 1990, 30).

Jayawardena's presidency marked the beginning of an aggressive development agenda, including market reforms, foreign investment, and export processing zones. These state efforts toward economic liberalization and de-welfarization—neoliberalization—needed to be legitimated by the implementation of massive rural development schemes, which went hand in hand with an expansion of the state budget and of the public sector (Venugopal 2018). The east would be central to this state expansion. The Accelerated Mahaweli Development Project (AMDP)—a two-billion-dollar irrigation and hydroelectric power scheme and Sri Lanka's biggest-ever development project—was the most ostentatious of these measures. As a modernizing project, the AMDP centralized state bureaucracy and privileged science and technology that was not just "compatible with but reincarnated" ancient indigenous Sinhala Buddhist national culture (Tennekoon 1988, 297). These state efforts also illustrate the long-standing ties between security and development in Sri Lanka. With each opening of a dam, the government performed elaborate Buddhist ceremonies and rituals, harkening to and reconstructing a glorified nationalist landscape of Sinhala hydraulic culture and civilization. If land and water could be imbued with Sinhala-ness, then Sinhala Buddhism could be considered inherent and "naturally" foundational to the island. By the time the cyclone struck in November 1978, Sri Lankan social and political life was already undergoing a gradual expansion of the state that was not inherently Sinhalese but that, as Venugopal (2018) characterizes it, became Sinhalized.

By the beginning of the war in July 1983, the east had already experienced a mix of ethnic and religious tensions, economic and political restructuring and class inequality, and state-sponsored colonization (see Spencer et al. 2014). The late 1980s witnessed the government violently put down the Janatha Vimukthi Peramuna (JVP, "People's Liberation Front"), a militant anti-state insurrection. The late 1980s also brought the violent departure of the Indian Peacekeeping Force (IPKF), who ended up battling with the LTTE—the LTTE allegedly armed by the Sri Lankan government. Over time both the Sri Lankan government and the LTTE demonstrated their violent power: the government's security apparatus was empowered by the existing State of Emergency and the Prevention of Terrorism Act, while the LTTE gained control of the north and the east after the Indian Army pulled out of Sri Lanka. Caught in the middle of these battles that included gross human rights violations on all sides, of course, were civilians. As Spencer et al. (2014) note, there are not many published accounts from the 1980s and 1990s, with a few exceptions (see Lawrence 1997). The University Teachers for Human Rights (Jaffna) bravely provided regular reports and bulletins, with sharp observations on the erosion of civil liberties of all parties swept up in conflict in the north and the east (see, for example, UTHR 1995).

In the east, the 1990s were particularly violent. Many of the people with whom I spoke about life in the east referred to the 1990s as the time of *prashanai* (problems). Until the Ceasefire Agreement between the LTTE and the Government of Sri Lanka was signed in 2002, entangled webs of violence constituted the eastern landscape; these included the LTTE's forcible recruitment of Tamil sons, STF roundups of young Tamil men, and territorial struggles between Tamils and Muslims, leading to the hardened fault lines referenced by Selvi in the beginning of this section. Over the years of the war, the territory of the east also became a space of displacement through occupation of land and homes by the military and LTTE, checkpoints, and High Security Zones (see Fonseka and Raheem 2010).

In December 2004, the tsunami crashed into the east coast, a region already coping with layered disasters. This year also witnessed the breakup of the LTTE's hold on the east, with eastern Tamil and former LTTE colonel Karuna Amman leading the charge. With his collusion, the Sri Lankan government was able to overtake the LTTE's control of the east; they proclaimed the region liberated in 2007. The government declared 2008 "The Year for War" and overpowered the Tigers by May 2009. In contrast with the cyclone of 1978—or rather, as I argue in this book, building from the cyclone—the tsunami engendered a new logic of disaster (that is, terrorism) management. As such, the next section focuses on the emergence of Sinhala Buddhist nationalist sentiment and institutionalization in national politics. The section offers a thread, highlighting political definitions of outsiders, aliens, terrorists, and the foundational structures to manage these so-called national threats that would, after the tsunami, become crucial to the government's technologies of disaster nationalism.

ALIENS, MINORITIES, TERRORISTS: SRI LANKA'S MODERN CONSTITUTION

As a construct, the "nation," or ideas of it, is ever shifting in relation to changing social forces. The nation, ultimately, is a homogenizing social project that affords certain groups of people power and legitimacy.[5] As Sivamohan Valluvan proposes, "Nationalism, as opposed to being a claim premised primarily on active belonging, is principally a wager of non-belonging, an assertion of the nation's 'constitutive outside'" (Valluvan 2019, 36). This section, then, looks to historical processes of nationalism in Sri Lanka after independence, in particular attending to how notions of ethnicity (Sinhala) and religion (Buddhism) were politically leveraged and increasingly institutionalized to define who or what could or would be excluded from the Sri Lankan national imagination. I outline these processes as some of the historical precedents that paved the way for Sri Lanka's approach to disaster and national security after the tsunami.

As scholars have cautioned against the overdetermination of ethnicity in understanding the broader political dynamics of the conflict (Spencer 2002; Venugopal 2011), I do not seek to instrumentalize ethnicity as an explanation of ideological nationalism. Rather, I seek to "thematize" it (Scott 2004) "around cultural-political interventions" to approach nationalism and ethnicity as historically shifting problem-spaces (Scott 2004, 18). Ethnicity was a way of conceptualizing difference in Sri Lanka, through the democratic relationality of concepts such as majority and minority—vestiges of British colonial constitutional experimentations; ethnicity as a tool of political mobilization for Sri Lanka's ruling elites was an instantiation of democratic ideals and reason (Hewage 2014), "laced with experiments in xenophobia" (Spencer 2008, 614).

For example, in the eighteenth and nineteenth centuries, the British recruited Tamil laborers from South India to migrate and labor on the coffee, tea, and rubber plantations. In population statistics for the British Crown Colonies, plantation laborers were categorized with merchants and other immigrants from India as "aliens and resident strangers" (Peebles 2001).[6] The effects of this categorization persisted over time, establishing the perception that "Indian Tamils" did not fit into the established identities of "indigenous" Ceylon Tamils and Sinhalese. In 1940, addressing the Jaffna Youth Congress, D. S. Senayayake, who would later become Ceylon's first prime minister, stated, "I am totally unconcerned as to which community an elected representative may belong so long as he is a member of the indigenous population. The Indian Tamils are not members of the indigenous population" (Russell 1982, 248).[7]

Despite establishing a community in colonial Ceylon for generations, after independence from the British in 1948 and establishing a democratic

Ceylonese state, "Indian" or Hill Country Tamils were not granted citizenship. Prior to 1947, India believed that Tamils in Ceylon should be considered citizens, while Ceylon believed that Indians should return to India. With this difference unresolved, the first act passed by the newly elected postindependence Parliament was the Ceylon Citizenship Act No. 18, by which a person could gain citizenship only by descent or by registration. To prove "descent" and hence citizenship, a person had to provide evidence that at least two generations (with preference given to the paternal side) were born in Ceylon. Citizenship by registration would be granted if an individual had resided in Ceylon for seven years, if married, or ten, if not married. Then, in 1949, the Indian and Pakistani Residents' (Citizenship) Act No. 3 was passed, also offering limited avenues of citizenship based on its stringent requirements (Jegathesan 2013; Kanapathipillai 2009). It was widely noted that even "indigenous" Sri Lankans, including even prominent political figures, would not be able to prove their citizenship according to these standards (Shastri 1999; UTHR 1995). Shortly after, also in 1949, the Parliamentary Elections Amendment Act made citizenship a requirement for voting, disenfranchising Indian immigrants and preventing them from running for elected governmental positions. Though Ceylon's early constitutional and parliamentary developments were generally forged along class lines and the maintenance of an elite class, these developments would also have deep and far racial ramifications and national exclusion for this group of "aliens" and "unassimilated minorities."

Combined with later constitutional policies, such as the Sinhala Only Act, which removed Tamil and English from the national language register in 1956 and land reform acts in the 1970s, Hill Country Tamils became an increasingly marginalized and mistreated population of people in Sri Lanka. Their citizenship and inclusion in national politics were routinely denied, their land rights were spurned, and their working and housing rights conditions were violated. To make matters even more difficult, with the rise of Sinhala and Tamil nationalist and militant movements during the 1970s and 1980s, Hill Country Tamils also bore the brunt of anti-Tamil violence and were regularly detained on suspicion of associating with or being LTTE (Bass 2008; Jegathesan 2015). The political and social exclusion and mistreatment of Hill Country Tamils, as aliens and "unassimilated minorities," is what Mythri Jegathesan (2013) refers to as their "heritage of dispossession" in Sri Lanka.

Detainments of Hill Country Tamils as suspected Tigers were enabled by the development of Sri Lanka's first anti-terrorism bill, "Proscribing of

Liberation Tigers of Tamil Eelam and Other Similar Organizations of Law," which was deemed an urgent bill and rushed through without public debate (Wickremasinghe 2010). It would eventually be replaced by the Prevention of Terrorism Act (PTA) of 1979. Though the 1979 PTA was introduced as a temporary measure, it was inscribed in 1982 as permanent legislation. The PTA was most certainly a response to and mechanism to clamp down on the Tamil insurgency, but it was also created to deal with the JVP (Janatha Vimukthi Peramuna) Marxist insurgency against the Sri Lankan state in the 1980s. These anti-state political insurgencies also motivated the ongoing state of emergency, which had been in place since 1971 with the first JVP uprising. The government justified the promulgation and use of the PTA and emergency rule by averring that the existing or "normal" laws were not adequate to ensure national security against ongoing threats (Uyangoda 2010).

In name, the Prevention of Terrorism Act in Sri Lanka was an attempt to mirror anti-terrorism legislation in the United Kingdom, but in practice it was actually much more "fearsomely draconian" (Lal 1994). The PTA is widely recognized and experienced as such, and human rights organizations in Sri Lanka and internationally have long averred its noncompliance with international human rights law and the fundamental rights granted by the Sri Lankan Constitution (Welikala 2008; Wigneswaran 2010). The PTA defines terrorism loosely as "any unlawful activity" and confers search, seizure, and detainment without warrant to police and state powers. During the war and even after its end, the PTA has focused on and targeted mainly Tamils, and more recently Muslims (a development I will further discuss in the conclusion), accruing a long list of human rights violations. According to the Colombo-based Centre for Policy Alternatives (2013), the PTA has "no place in a democratic society" and "fl[ies] in the face of almost every conceivable human rights norm pertaining to the liberty of the person." The long list of offences of the PTA include (but are not limited to) detention without charge or trial for extended periods of time in irregular places of detention; violation of detainees' rights, including through torture and physical, emotional, and sexual abuse; the shifting of evidential burden of proof to the defendant; and disproportionate penalties. The PTA also restricts freedom of expression of the media, sometimes requiring government permission to print, publish, and distribute publications and newspapers.[8] As Vinay Lal claims, "the infliction of 20 years imprisonment for defacement of public notices [is] nothing short of being fascist" (Lal 1994).

Today, over a decade after the end of the war, the PTA still remains in place, enabling executive forms of power that were previously upheld by the state of emergency, which was finally lifted in 2011, two years after the end of the war. This, despite the fact that in October 2015, the newly elected president Maithripala Sirisena promised to repeal the PTA. After the war's end, many members of civil society, human rights lawyers, activists, and organizations had called for a repeal of the Prevention of Terrorism Act. Upon ousting Mahinda Rajapaksa in 2015, Sirisena claimed that he would work toward this. But in 2018, while Sirisena was still president, Parliament drafted a new counterterrorism law that was widely criticized as even more draconian than the PTA. Of foremost concern was that, like the Disaster Management Act, the bill defined "terrorism" capaciously.[9] This law did not pass. Recently ousted president Gotabaya Rajapaksa added a new set of regulations, The Prevention of Terrorist (Proscription of Extremist Organizations) Regulations No. 2 of 2021, which the International Commission of Jurists (2021a, 2021b) condemned for being ill-defined and abusive. These regulations came on the tail of a set of PTA regulations on "de-radicalisation," which, building on established regulations in the original PTA of 1979, allows for arbitrary detention for up to two years without trial.

THE NATION HAS NEVER BEEN HOMOGENEOUS

With this history of state-sanctioned exclusion and the creation of powerful policies to regulate and "protect" the nation from terrorism in mind, I now return to contemporary disaster risk management—the logics and practices of disaster nationalism—in Sri Lanka. In particular, I have highlighted the ways in which Sri Lanka's constitution, postindependence, has protected, enabled, and upheld Sinhala Buddhist nationalist hegemonies and fantasies, to the exclusion and detriment of ethnic and class minorities. Key to this evolution of state power is how, historically, Sinhala nationalism and state-building has defined and categorized "minorities" as less than or not Sri Lankan and, ultimately, as external "threats" to the nation. That is, the prevailing logic of national security and securitization in Sri Lanka is premised upon its categorization of marginalized others, who, in different historical and political moments, have been referred to as "aliens," "terrorists," and "foreign invaders." These codified "outsiders" of the nation, denied the rights of Sri Lankan citizens, are dehumanized and cast—along with nature—as threats to the imagined pure integrity of the nation. Disaster management, then, as the governance of natural disasters and terrorists, emerges as a mechanism of national securitization. What we learn through this history and in my examination of the east and the cyclone, however, is that even with this precedent of exclusion, disasters were not (yet) operationalized in service of state-sponsored nationalism.

Here, the entanglement of natural disaster and terrorism by the Sri Lankan government is not a celebratory ontological political move. Rather, as Sarah Ives (2019) reminds us, the blurring of humans and non-humans is also part of violent racial and colonial histories in which the dehumanization of "inferior" humans has long been a practice of power (de la Cadena 2011; Mavhunga 2011)—especially in war and strategies of counterterrorism (Bhan and Bose 2020). In this, Sri Lanka is not alone. For example, terrorist detainees in Guantanamo Bay are "excluded from the range of rights that constitute the legal category of the human subject" (Pugliese 2013, 107; see also Butler 2004), often characterized as "monsters" (Pinfari 2019) or "vermin" (Steuter and Wills 2010). The abstraction of humans, even children, as "threats" is central to logics and practices of national security (Shalhoub-Kevorkian 2014). In Sri Lanka then, the figure of the terrorist, with its juridical status outside humanity, is excluded from the realm of the political or of the nation and thus, the terrorist or even its specter is abstracted into the realm of the "nonhuman."

The designation of terrorists and otherwise threatening "others" as nonhuman provides an understanding, then, of Sri Lanka's contemporary preemptive approach to disaster and disaster risk. I return to the speech delivered by Sri Lankan Minister of Disaster Management and Human Rights Mahinda Samarasinghe at an international conference on Disaster Risk Reduction in Geneva just after the end of the war in 2009:

> My country—Sri Lanka—has just overcome a human-made disaster of a magnitude unparalleled by any similar recent events elsewhere. We have overcome the scourge of terrorism that has beset our island nation for well over two decades. . . .
>
> All our efforts at renewal, rebuilding and resettlement, however will be put at risk if the cause factors of the conflict and terrorism are not addressed and our President has committed himself to evolving a home grown political response to those factors.
>
> Borrowing from DRR methodology, our political response will reduce the risk of a renewed human-made disaster, i.e. terrorism and

conflict, through systematic efforts to analyse and manage the causal factors, evolve consensual responses and improved preparedness for adverse events. We do not for a moment think that Sri Lanka's national renewal will be quick or easy.

There are ever-present threats that we must, and will, guard against, including the threats of new violence and destabilization.

Here the horizon of disaster stretches into an imminent, interminable future, justifying ongoing practices of securitization and militarization in an always vulnerable Sri Lanka, further perpetuating the anxieties and precarity of disaster-worn communities across Sri Lanka and especially in the east.

In working toward Sinhala Buddhist nationalist purification, the Sri Lankan state simultaneously creates the threats to its own homogenizing fantasies. Recall Rajapaksa's victory speeches that identify the "enemies" of the "peaceful" nation. The work of national purification in Sri Lanka redoubles the vexed and problematic social forms and imaginaries of modernity. In their research on post-tsunami experiences in the east, Shahul Hasbullah and Benedikt Korf (2009) argue that disasters provide opportunities to reimagine the state as a purification mechanism—operating through methods of control, bureaucracy, and regulation in post-tsunami housing reconstruction (see also Korf 2006b). As the remaining chapters of this book will detail, these modes of purification—fantasies of disaster nationalism—are never fully realized. Indeed, the work of national purification, as Jonathan Spencer (2003) draws from Bruno Latour, is a modern ideal, with a far messier reality: "Official control over movement is never complete, and frequently quite ineffective, yet it remains a privileged site for the state, through its agents, to display a certain vision of political order" (2003, 21).[10] The nation-state as a particularly modern political phenomenon and imagination thus invokes parallel processes: even as state power works to purify the nation through practices of ordering, accounting, regulation, and control, these efforts are made ever more implausible due to the violences, separations, and exclusions of these very practices (Visvanathan 1997). The nation, even at its most cohesive, is fragmented (Chatterjee 1993). The nation has never been pure. The nation has never been homogeneous. The nation-state, thus, is not a given, but is in fact formed through these shortcomings.

As Joseph Masco observes, today terror is "an open-ended concept, one that links hugely diverse kinds of threats (nuclear, chemical, and biological weapons to be sure, but also attacks on the public image of the United States, computer hacking, infectious disease, and disruptions to daily life, to name but a few) and treats them all as equally imminent, equally catastrophic" (Masco 2014, 19). We witness a similar type of open-endedness with Sri Lanka's Disaster Management Act and an unsettling expansion of the security state: "counterterror sets no conceptual or territorial limit to defense" (Masco 2014, 19). As Laleh Khalili delineates, it is the *failure* of colonial counterinsurgency technologies—in her research, in Occupied Palestine—that makes counterinsurgency a self-sustaining and self-justifying mechanism requiring "ever-more sophisticated modes of control" and "ever-expanding populations of suspect civilians" (2010, 427).

Between 2008 and 2009, during the height of the Sri Lankan government's military campaign against the LTTE, dengue fever had reached unprecedented levels. (Presently, the disease continues to elude eradication efforts). After the end of the war, walking around Colombo where dengue cases had been particularly prevalent, I saw a poster that must have been part of the Ministry of Health's dengue campaign. I did not take a picture; I did not have my camera on me, and these were the days before the ubiquity of smart phones. However, I will never forget what the poster read: "Just as we eradicated terrorism, we will eradicate dengue!" While I failed to find a copy of that handbill, the Sri Lankan Army documented their effort "to go all out on Dengue eradication" in their online news section in October of 2009: "The soldiers who eradicated terrorism from the country through offensive operations have this time lined up to launch an all out offensive against Dengue outbreak on a concept of the Commander of the Army Lieutenant General Jagath Jayasuriya" (Sri Lanka Army, n.d.).

Here too, a mechanism of purification was at play.[11] Terror became so naturalized in the national imagination that even the nation's health would be militarized.[12] Police wandered around neighborhoods, ticketing and fining homes with "dirty" exteriors where mosquitoes could potentially breed. I arrived back to my host family's home one afternoon to find everyone scurrying around outside picking up leaves and brush and moving plastic furniture and other castaways that had accumulated outside the home. They had been warned by the police that if they did not tidy up their home exterior, they could create more mosquito breeding grounds and, moreover, would be fined. Using the language and imagery of immunology, Emily Martin shows how the body as a nation-state is always potentially under attack, rhetoric that naturalizes and normalizes violence: "As immunology describes it, bodies are imperiled nations continuously at war to quell alien invaders. These nations have sharply defined borders in space, which are constantly besieged and threatened. In their interiors there is great concern over the purity of the population—over who is a bona fide citizen and who may be carrying false papers. False intruders intend only destruction, and they are meted out only swift death. All this is written into 'nature' at the level of the cell" (Martin 1990, 421).[13]

Rajapaksa's narration of the nation persists, but, as the rest of the chapters show, these designated threats to the nation are not the only threats that Sri Lankans must negotiate.

Even the blue calm reminds you there was a before, though it promises no safe after.

V. V. Ganeshananthan

Tsunami memorial, Batticaloa, 2017. Photo by author.

2 Anticipation

Wiping the sweat from my brow, I switched my gaze from the tsunami warning tower peeking over the tree line—and beyond, where the ocean lapped on the eastern shores of the island—to the quiet, empty street, lined with the houses of locals who would be "evacuating" their homes once the warning siren went off. They had been given advance notice by the local disaster management coordinator that they had been selected to participate in a government-mandated national tsunami evacuation drill. The din of drums and music at a nearby temple festival could be heard amid the silence of waiting. For it was July, also the season for the annual Kataragama festival and the Pada Yatra. The Pada Yatra is the annual walking pilgrimage that starts in May in northeastern Sri Lanka and ends in July in Kataragama, a city in southeastern Sri Lanka famed for its multifaith religious temples and festivals. The Chitra Velayudha Swami Temple in Thirukkovil, whose festivities we could hear now, is a resting point for those making their way to Kataragama.

I looked at my watch: 3 p.m.

I looked toward the tower: silence.

Back to my watch again: 3:05 p.m. Still nothing.

At 3:15 p.m., I sent an SMS to my contact at the district disaster management office: "What's wrong?" I asked.

"Just wait," he texted.

So we waited.

And still: no siren.

I stood there waiting for the action to begin. How loud would the siren be on a day like today, with no wind, when the siren could supposedly be heard throughout a five-kilometer radius?

I turned back to the street behind me, which was no longer empty. It was slowly becoming populated by people trickling out of their houses with their "emergency bags." Some brought out plastic chairs so that they could sit in front of their homes while they waited for danger to be signaled.

I switched my gaze back to the tower.

We were all ready.

Finally, at around 3:30 p.m., the local disaster officer answered his cell phone, nodding his head and appearing to take directions. Upon hanging up, he quickly picked up his bullhorn, climbed atop his colleague's motorbike, turned on the decidedly less-dramatic-sounding bullhorn siren, and raced down the street. Understanding this cue, the designated "evacuees" began their movements toward the tsunami "safe site"—a small Hindu temple about one kilometer inland. Some ran, shouting, carrying their disaster bags, acting as if danger lurked nearby. Others strolled leisurely, laughing and talking along the way.

Following their commitment to Disaster Risk Reduction and as outlined in the "Roadmap towards a Safer Sri Lanka," the Disaster Management Centre and the Sri Lankan state implemented new forms and spaces of knowledge- and security-building. One major project set up a national early warning system for natural disasters. Initially, the warning system consisted of the construction of multi-hazard warning towers dotting the Sri Lankan coastline—a total of fifty in all, with twenty-five more slated to be built in the newly "freed" northern area part of the island.[1] The towers are all connected; they communicate with each other via satellite from the village, to the district, and all the way up to the national level. The main operating center functions out of the Disaster Management Centre located in the capital city, Colombo. The towers deliver warning messages to the coastal communities that live around them. According to the Disaster Management Centre, the sirens can be heard within a five-kilometer radius. The Department of Meteorology in Colombo is first to get word of an earthquake or a potential tsunami from their information sources—namely, the Japanese Meteorological Association and the Pacific Tsunami Warning Center in Hawai'i. They then contact the operating center at the Disaster Management Centre, which can either directly send a message to the towers to signal a warning or instruct district-level coordinators to manually turn the warning sirens on. The messages are recordings, spoken first in Sinhala, then in Tamil, and finally in English. The first message issues a warning. If evacuation is necessary, then an evacuation message will be broadcast. Finally, when the danger passes or there is no danger, a cancellation message will be issued.

To test the system, the government staged several tsunami evacuation drills. So on July 20, 2009, I found myself staring up into a cloudy blue sky to observe the technology and warning system in action. My friend, a disaster coordinator working with the United Nations Development Program (UNDP) and the Disaster Management Centre, had invited me to the drill. I wiped the sweat from my brow and shifted my feet standing in the muggy, hot, late afternoon sun. We turned and faced east, toward the ocean, although tall trees and homes obstructed our view of it. Behind those trees and homes stood a newly erected disaster warning tower. This new one, located on Sri Lanka's southeastern coast, stood above rooftops, bright and shiny, red and white (see figure 2.1).

This chapter considers preparedness and preemptive modes of disaster governance concretely, by exploring the temporal modalities of anticipation in disaster risk management and governance in Sri Lanka. In particular, it follows the logic of disaster risk management and control, which is structured upon and oriented toward an *anticipation* of disasters, the homogeneous time of disaster nationalism (Choi 2015). Anticipation might evoke a familiar structure of feeling in our contemporary security moment and in the looming era of the Anthropocene (see Cons 2018; Paprocki 2019). Hinged on a notion of futurity, and in particular a future rife with risk and insecurity (Beck 1992), anticipation also works as an apt temporal heuristic to examine new modes and logics of governance, security, and subjectivity (Adams, Murphy, and Clarke 2009; Anderson 2010; Anderson and Adey 2012; see also Tironi, Rodriguez-Giralt, and Guggenheim 2014). Literature in anthropology and science and technology studies on risk illustrates as much, focusing on the practices of experts and expertise, technoscience, and state-based

FIGURE 2.1. Tsunami warning tower. Photo by author.

securitization that deal with the production of anticipatory knowledge and practices—from the speculative life sciences (Fortun 2006), to health (Dumit 2012), to Wall Street (Zaloom 2004), to imagined pandemics (Lowe 2010; Porter 2019) and biothreats (Keck 2020; Lakoff 2008).[2] Indeed, as briefly explained in the introduction, Sri Lanka's risk management and anticipatory governance result in new preparedness and preemptive actions (Cooper 2006; Lakoff and Collier 2015; Massumi 2005; Stewart, n.d.; see also Foucault 2007), issuing new technical Band-Aids such as warning systems, infrastructure, evacuation drills, and event simulations as well as overall attempts to increase government management and control. The impetus of such programs and collaborations is to invoke a continual state of readiness and maximum security of state territory: it is not a matter of *if* a disaster strikes, but a matter of *when*. Anticipation, thus, is productive. It reorganizes and restructures lives, nations, and economies; it generates new biosocialities, ethical plateaus (Fischer 2003), technologies, and orientations to the future and to the past. Anticipation is profoundly political, even "psychopolitical" (Orr 2006), bringing together new regimes of fear, hope, and governance (Masco 2008).

In Sri Lanka, the anticipation of disaster has proven to be no less generative in restructuring the nation. Recent scholarship on regimes of preparedness and biosecurity—or what has been termed "vital systems security"—pinpoints infrastructure and territory as what needs to be maximally securitized by the state (Lakoff and Collier 2015). In Sri Lanka, however, while certain war-related infrastructures were in place—the checkpoint being exemplary in this regard—other everyday, practical infrastructures had long been neglected or affected by the war.[3] Sri Lanka therefore did not have an ideal state of "normalcy" to restore after the tsunami. War had long scarred the social and political landscape. As one development official pointed out, in the war-affected regions of northeastern Sri Lanka, it was difficult to decipher if some of the destruction was caused by the tsunami or war-related bombing. The Sri Lankan state's technologies of disaster nationalism aimed not only to protect and maintain existing infrastructures but also to build new ones.[4] The ruins of the past would have to give way to a new vision of the nation: a Sri Lankan nation prepared for disaster and, finally, freed from the scourge of terror. Buffer and border zones, national disaster early warning systems, military checkpoints—these state-employed technologies of control cemented the construction of the newly liberated "undivided Motherland."[5]

Yet, these governmental attempts to control national futures remain in constant tension with the attitudes and opinions of people who have been affected by both the tsunami and war. These collective relations, practices, and infrastructures of feelings are what I refer to as "anticipatory states." My use of the plural "states" is deliberate, to acknowledge the ways in which anticipation has multiple valences and resonances in Sri Lanka. From the calculative risk management projects and anticipatory practices of the Sri Lankan state to the everyday state of being ready and aware in the spaces of disaster, anticipation weaves into and out of experiences and encounters, its different forms and possibilities shaped by complexly layered histories and landscapes of disaster and violence. For the people I knew living in eastern Sri Lanka, the possibility of a disastrous future depended upon what they knew and had experienced in the past. For while an anticipatory logic of preparedness conjures up an imagination of a catastrophic future, for them, the impossible, the unimaginable, had already happened. As the rest of this chapter shows, their anticipatory actions critique and illustrate the limits of the anticipatory Sri Lankan state. The unified and securitized Sri Lankan nation as the medium of state power exists as a "fragile equilibrium" (Aretxaga 2003), an order that is susceptible to disruption, when a place, person, or event remains unaccounted for in its control and order. As the Honorable Minister Samarasinghe pointed out in his speech in Geneva, the always-dangerous future that must be harnessed to maintain peace is crucial to the imaginary of the nation. As we will see, the fantasy of total control is susceptible not just to the threat of disaster but also to its own impossibility—a Nervous System indeed (Taussig 1992).

ANTICIPATORY STATES

As previously mentioned, I lived in Sri Lanka during the government's aggressive military campaign between 2008 and 2009. When the war came to its dramatic end in May 2009, I was in the bustling coastal town of Kalmunai, in the Eastern Province. I was there to learn about the experiences of Sri Lankans whose lives had been impacted by both the tsunami and the civil war. As detailed in chapter 1, as Sri Lanka's "crucible of conflict" (McGilvray 2008), the east is characterized by complexly layered social and political histories of violence, migration, displacement, and resettlement. For many years, and especially during the 1990s, the Eastern Province was much fought-over territory; it was considered part of the LTTE's imagined homeland until 2007, when Sri Lankan government forces, assisted by LTTE breakaway leader Colonel Karuna Amman, "liberated" the region.[6] And remember: the east coast was the area hardest hit by the tsunami. It was the part of the island located closest to the epicenter of the tsunami-generated earthquake near Aceh while also being the most densely populated area on the island. In this section I bring together the different ways in which anticipation, as a logic of state power crucial to processes of disaster nationalism in Sri Lanka, takes shape in this disaster-stricken area. Preserving and upholding Sri Lanka's newfound freedom enabled the moral justification of increased militarism and securitization of areas of Sri Lanka after the tsunami and after the war, in turn leading to a palpable lack of social and political change.[7]

The evacuation drill took place two months after the government's defeat of the Tigers. Yet the end of the war felt far from settled. At the time, hundreds of thousands of Tamil civilians were still detained in government-guarded internment camps and security concerns across the island were still high. In these uneasy times, I found myself participating in the failed tsunami evacuation drill with which I began this chapter. My disaster-coordinator friend who had invited me to the national drill, rather presciently, joked with me as we were getting tea and snacks that the sustenance would come in handy if the drill was a failure and we had to run from people who were upset with the government.[8]

For this was not the first time the siren had failed to sound, either as scheduled or during a real tsunami scare. In 2007, a series of megathrust earthquakes rocked the Sunda Trench off the coast of Indonesia, causing several tsunami bulletins to be issued. At that time, the Sri Lankan government had erected three pilot warning towers. In a different eastern coastal town, one of those towers stood steadfast and ... silent. My friend, the local disaster coordinator of that area, recounted to me that repeated attempts to contact the Disaster Management Centre in Colombo failed. People started leaving their homes. Military personnel came around to usher people away from the coastline. Finally, my friend told me that he climbed the warning tower himself and manually turned the siren on. By that point, nearly all the people in the surrounding areas had already evacuated inland and to higher ground.

At the Disaster Management Centre in Colombo, I was told that the 2007 incident was a success. An official informed me that he was "proud" to say that "without harming [sic] to any single human, we [had] evacuated people in coastal areas within 30 minutes." He prefaced this statement by saying that Sri Lanka was "lucky," as the event was "good practice." The role of the Disaster Management Centre (DMC), as he further explained to me, was to get correct details and accurate information. During the time of the tsunami and the tsunami scare, the DMC was dependent upon information from other sources—as it continues to be to the present day. The DMC collaborates with the Department of Meteorology to monitor for tsunamis and cyclones, while the Irrigation Department monitors for floods. The official went on to tell me, "We must give correct information and then let people know who are affected." Accordingly, he told me that the DMC now had "confidence about people; now they will not react without the proper information from the DMC."

"Yes, I received a call from my brother in Trincomalee [a town on Sri Lanka's northeastern coast] that a tsunami was coming," Fathima relayed to me. Her brother, a fisherman, had been listening to the radio and heard some news about an earthquake in Indonesia, which immediately prompted him to call his sister, who lives in the area that was most affected by the tsunami of 2004. Fathima's home is roughly ten meters from the warning tower. She said she did not wait to hear it ring, but instead gathered up her family members, made phone calls to relatives and close friends, and spread the word to her neighbors. She said that in a matter of a half an hour, nearly everyone in her village had begun the journey inland toward the main road.

Feelings toward these towers were ambivalent. As I sipped a warm Coke near the cooling breeze of the ocean, Mohamed, a local owner of a small *kadai* (shop) on the beach, relayed to me that he was happy about the tower. In fact, because of the warning tower, he had felt comfortable enough to reopen his shop near the beach. "At night I can sleep better," he said, looking toward the expanse of ocean. He blinked slowly, "Before I could only listen to the sound of the waves, listening for the way the waves sounded the day of the tsunami." Still, others were skeptical: at a local tea stall, near the same tower to which Mohamed referred, I asked some villagers what they thought. Raheem took a sip of his tea, smiled, and told me, "It is there." On visiting other towers in other coastal regions of the island, people informed me that the warning towers had sometimes gone off at unscheduled times, occasionally inciting moments of fear and panic. Lakshmi informed me: "We are always alert." She, like many others living near the shoreline, takes note of the slightest shift in wind, the color of the sky, and the behavior of birds. The day of the tsunami, the sky was a gray color and engulfed with clouds. That gray sky was also remembered by my friend Hafeez, who recalled its somberness when he rode his motorbike to work. On cloudy days, Lakshmi relayed, she often felt nervous and avoided going to the sea, like many others (see also Choy and Zee 2015). Sometimes false alarms and rumors of tsunamis spread on days with high winds. On those days, many fishermen would stay home and avoid going out to the rough sea. Temple and mosque loudspeakers now double as potential community warning broadcasters. Alongside and beyond the preparedness rationale of the government, people too consider the possibility of another disaster, using their own modes and social networks of communication to get critical information.

But life in the east required more than being prepared for the next tsunami. Despite the end of the war, especially on the east coast, military presence continued unabated. Although the area had been "freed" from so-called terrorism in 2007, an official "state of emergency" remained in place.[9] The state of emergency granted military-like powers to the "Special Task Force" (STF), an arm of the Sri Lankan police. In those months following the proclaimed end of the war, military checkpoints, like the warning towers, dotted the physical landscape of the east. These checkpoints were located on main roads, stopping cars and other vehicles to check for bombs and hidden terrorists. Gun-wielding military personnel checked IDs and bags, and sometimes patted people down. The presence of these checkpoints constituted an anticipation of violence, the possibility of a bombing: the "not-so-diffuse" (Jeganathan 2000) tentacles of the state. The STF continued to monitor roadblocks and checkpoints, while also conducting neighborhood and house checks, making sure that random strangers—or potential terrorists—were not taking refuge in the area. The officers could be seen roaming the streets, entering homes, and lounging at local tea stalls. Indeed, as an interview with the Commanding Officer of the STF of the eastern district revealed to me, "The war is over, but terror is there." He continued to inform me that the division I was researching, and in which I was living, was deemed a "vulnerable" area—an area vulnerable to lurking terrorists—so it only made sense to increase the number of troops as a measure to counter terrorism and "secure the situation." A couple weeks after this conversation, the Sri Lankan Army began rebuilding an old military camp that had been run down for more than a decade. These anticipatory practices of everyday military occupation and national vigilance focused on purifying these vulnerable and potentially radicalized spaces work to systematically produce violent spatial and affective control (see Junaid 2013).

With the ever-present STF conducting their checks in the east, my friend Parvati explained to me how she stayed ready and alert: "As long as we have all our documents and have registered all of our family members with the police, the STF will not give us any trouble when they are doing their household checks." The fear of arrest or violence remained palpable after the end of the war, when people would confess to me in hushed tones their disbelief that the LTTE leader Prabhakaran was really dead. A rumor went around about a boy on a bus who, unaware of the presence of a police officer, proclaimed his disbelief regarding Prabhakaran's death and was unexpectedly smacked and threatened with arrest for making such a statement. My friends could only tell me, with resignation, that, "inside, we feel sad." There was a lingering sense that the resumption of war was still possible.

In fact, people commented on my seeming naïveté when I probed them about the end of the war. "Who told you the war was over?" asked Ravi, a young Tamil man. Ravi was home for just three months, on holiday from his job in Qatar. When we were talking, I mentioned that his visit had coincided with a historic moment in Sri Lanka's history. I asked him, then, if he would stay in Sri Lanka, and return for good from Qatar? He said his plan was to go back to Qatar, perhaps to return to Sri Lanka if he got married and had a family. For now, he informed me, he felt that the east remained unsafe. He further explained, "The government says the war has stopped, but there is no full freedom for people to roam and move about freely. When that happens, we can say that the war is finished." After hearing about my research project, he remarked, "If you release a fact about what is happening here that makes the government unhappy, the next day you can't come back here. If there's no safety for foreigners, how can there be safety for locals?" Fawzi, who was living in a block of post-tsunami flats in the Muslim division of town, challenged me: "Have the food prices gone down? Are there new jobs? If they [the government and the LTTE] want to start up the war again whenever they want, they can." Sitamma explained the demise of the LTTE to me like this: "In the sea there is a lot of fish, but can you catch all the fish? Any little thing can start things up all over again. And if they [the government and the LTTE] want to start something again, they can."

So after the war's declared end in May 2009, no easy peace had come to settle over Sri Lanka. The rebuilding of the nation requires a constant vigilance, and the past hovers uneasily over daily life. How do realities and experiences merge and emerge amid national orderings, out of the past and into the future? Here, I offer that vigilance is not merely a technique of self-disciplining for those living near the shorelines and in the spaces harboring "terrorism." Living in eastern Sri Lanka requires a type of anticipation that sees neither the past nor the future as closed (Bloch 1995; see also Jameson 2002). This anticipatory state is a conscientious recognition and critique of the contradictions of the material and social conditions and, more crucially, the legacies of state power, in and through which many must continue to live. This anticipatory state enables people to go about their daily lives—carrying worry with them, but not being debilitated by it. The people I came to know, having experienced tsunami and war and having lived in and with these complex and precarious intersections of objects, spaces, persons, and practices, created an anticipatory platform for living through and enduring times when the possibility of another disaster always looms.[10] Anticipation, for many exposed to the ongoing insecurities of disaster management, is a recognition that the "ends" of disasters lead to neither a stable peace nor emotive complacency (on the "ends" of war, see Nelson 2009; Nelson and McAllister 2013).

ANXIOUS STATES

Yet as state power has become a fixture in everyday life, it is also a Nervous System (Taussig 1992) constantly undermined by its own fragility. Returning to the tsunami evacuation drill: the day we stood under the hot late-afternoon sun in July waiting for the alarm bell to sound was special. The drill had been planned in relation to and ahead of a major and eagerly anticipated astrological event. Two days after the drill, the longest total solar eclipse of the twenty-first century would occur. Excited scientists and observers from the world over converged around different parts of Asia, where this once in a lifetime total eclipse would be visible. Tour groups and even cruise ships were chartered to catch the best glimpse—the totality of the eclipse—from the middle of the Pacific Ocean. Astrologers in Sri Lanka and beyond also recognized the exceptionality of the eclipse. Some predictions suggested that following the eclipse, great catastrophes could befall the world—from catastrophic natural disasters to another world war. The predictions were printed in the local newspaper, and people were on high alert. Rumors of another tsunami began to swirl before the July 22 solar eclipse.

My good friend who worked for the local disaster management office forwarded me an email that had been circulating from a "tsunami quack," as he called him:

> Hello there. I just wanted 2 let you know that please stay away from the beaches all around in the month of July. There is a prediction that there will be another tsunami hitting on July 22nd. It is also when there will be sun eclipse. Predicted that it is going 2 be really bad and countries like Malaysia (Sabah & Sarawak), Singapore, Maldives, Australia, Mauritius, Sri Lanka, India, Indonesia, Philippines are going 2 be badly hit. Please try and stay away from the beaches in July. Better 2 be safe than sorry. Please pass the word around. Please also pray for all beings.

In the email, a map detailed the areas that were in the path of the eclipse and would therefore be in the path of another earthquake-induced tsunami. The source, according to the popular urban-myth-busting site Snopes.com, was a blog post positing that the gravitational pull would be so great during the eclipse that it would cause a seam to burst in the tectonic plates near Japan, creating a tsunami that would then affect the countries near Japan and across the Indian Ocean.

However, as scientists have pointed out, earthquakes are not caused by gravitational pulls. They are caused by shifts in the tectonic plates—the earth's crust. Moreover, if it were possible to predict tsunamis by eclipses, earthquakes would also be predictable; the reality according to scientists is that predicting earthquakes is an impossible task.

Still, the Sri Lankan government could not ignore the hype. Impending disaster forced disaster management into action. In the *Daily News*, one of Sri Lanka's English-language daily newspapers, an article came out two days before the eclipse, the day the tsunami warning drill was conducted:

> The Meteorology Department categorically denied the rumour that a tsunami will hit Sri Lanka, Singapore and Malaysia on July 22. Many people in the coastal areas had panicked by the rumour which has no scientific basis.
>
> There has been a rumour with a solar eclipse would trigger a tsunami affecting several countries including Sri Lanka on July 22, which has no scientific basis whatsoever, Director General, Meteorology Department, C. L. Chandrasena said.
>
> He said they are constantly on alert, 24 hours, to inform the public through the media and other sources in the event of a tsunami, cyclone

or any other disaster situation. Therefore, he had asked the public not to be misled by any rumour from un-official sources, he said.

Chandrasena said that the Sri Lanka Met Department, Geological and Mines Bureau and the Disaster Management Centre are the official bodies who inform the public on the possibility of a natural disaster in the country. (Senewiratne 2009)

While the article does say there is no "scientific basis" for the rumor, they still decided to shift the date of the evacuation drill and carry it out, *as if* there were something to actually fear and to allay fears and anxieties being expressed, especially by coastal populations.

Astrology is a social force that cannot be ignored in Sri Lanka, even by the government. It is put to use in everyday practices from marriage pairings to auspicious dates for travel and weddings. Even politics is steeped in this tradition; all political figures consult with their personal astrologers, seeking appropriate days to make speeches and travel plans, including then president Mahinda Rajapaksa. He took astrology so seriously that in June 2009, about one month after the war was declared over, authorities arrested one of Sri Lanka's most popular astrologers, Chandrasiri Bandara, because he predicted that President Rajapaksa would be ousted from office on September 9, 2009. This did not sit well with the administration. The president's spokesperson, Lucien Rajakarunanayake, stated, "We have to wonder why an astrologer would say such a thing. In Sri Lanka, astrologers are not just for fun. They play powerful roles in steering outcomes. *Saying it can make it so*" (Wax 2009, my emphasis). Upon his arrest, the "opposition" astrologer Bandara responded, "The Sri Lankan Government arrested and imprisoned me, but they could not prison [*sic*] Saturn" (Wax 2009).

Bandara's is one among many stories of the arrest, death, or abduction of those dissenting from or challenging the Sinhalese nationalist party lines. From 2006 to 2014, under the Mahinda Rajapaksa regime, Reporters without Borders recognized Sri Lanka as having one of the worst press freedom records in the world. The organization also highlighted the nation as one of the most dangerous places in the world to be a journalist, where dissent or government criticism were met with arrest, kidnapping, assault, and even death.[11]

Many of these kidnappings and murders are associated with "white van culture" in Sri Lanka. Unmarked white vans occupy a particular place in the Sri Lankan social psyche as a symbol of danger and secrecy. Victims, such as journalists or critics of the government, are picked up in white vans, sometimes never to be seen again or ultimately found dead or tortured. In 2009, as the war raged in the north, it was rumored among foreign aid workers that the Criminal Investigation Department (CID), Sri Lanka's CIA, had set up a hotline to report the "suspicious" behavior of foreigners. Signs went up all over Colombo about the ways in which foreigners who did not support the government's full-force efforts in the north were "aiding and abetting" terrorism through such actions. Dissent, or even moderation, was not tolerated. I witnessed several of my expatriate acquaintances and friends forced to leave the country because the government did not renew their visas.

Even two years after the end of the war, on April 17, 2011, an article in Sri Lanka's *Sunday Times* newspaper stated: "Legal action will be taken against astrologers, academics or others who make predictions on natural disasters and thereby cause panic among the people, Disaster Management Minister Mahinda Amaraweera warned yesterday" (Wickremasekare 2011). The police were given powers to arrest any individuals causing panic among the people. The government was concerned by the "scaremongering" that emerged from a set of predictions made by a geology professor at the Sri Lankan University of Peradeniya and his team of researchers. The team had predicted a series of earthquakes in the Indonesian region between April 10 and 18. As this was holiday season (the New Year) in Sri Lanka, many had traveled to the south and to the beaches, places that had been badly affected by the tsunami. According to the same article, because of the predictions that had been cast, many holiday-goers (including then president Rajapaksa and his family) were dogged with worry and concern.

The popular Sri Lankan independent online news organization Groundviews.org conducted a brief interview with the geologist, Atula Senaratne, who provided some context for his prediction: "Many people have been asking me if there is a basis for a Mayan warning suggesting the world could end in 2012 with a geological catastrophe. There is no current capability in geoscience to predict this, but this made us curious. So my team and I looked up the historical data on all the major earthquakes, and in particular those that have occurred during the past 100 years" (Gunawardene 2011). The gist of Senaratne's inquiry is that in the historical data, there appears to be a certain alignment of the sun, moon, and planets in approximately 75 percent of the 136 significant earthquakes over the last hundred years. While this is not necessarily conclusive, Senaratne said it was significant enough to pursue. In doing so, he and his team made further predictions to test their theory. The first prediction was made for February 20, 21, and 22, and then the earthquake in Christchurch, New Zealand, occurred. Senaratne pushed further and predicted earthquakes in the Pacific Basin between March 3 and 10. On March 11, an earthquake hit Sendai, Japan, triggering a devastating tsunami and what would become known as the "triple disaster." They made more forecasts about April earthquakes along Turkey and the border of India-China, which, according to USGS records, also happened. As their predictions "came true one by one," Senaratne thought this would be of public interest and started engaging with the media.

Senaratne's intent was not to stir up panic, he said, but rather to make information public on a topic that could potentially be of interest and use to a broader concerned audience. Scientifically speaking, he was attempting to determine whether gravitational or magnetic connections between celestial bodies might potentially yield earthquake activity. Senaratne's interest in planetary sciences is part of an emerging science called astrogeophysics, and in his interview he made sure to reiterate that his work "has nothing to do with astrology! We are using the scientific method." The only legitimate and publicly permissible form of scaremongering was that generated by the Sri Lankan government.[12]

ENDURING DISASTERS

The political cartoon in figure 2.2 appeared in the Sri Lankan English-language newspaper the *Daily Mirror* on April 12, 2012—almost eight years *after* the tsunami and nearly three years *after* the "end" of the war. The cartoon's publication followed a tsunami scare that had occurred just the day before. An earthquake with a magnitude of 8.6 struck off the coast of Aceh, inducing a series of tsunami warnings and general panic for countries across the Indian Ocean basin once again. In Sri Lanka, an official warning was issued too late by the Meteorology Department; it was issued *after* the time the tsunami would have hit Sri Lanka's shorelines. Even still, as my friends and news outlets reported, many, including other government agencies, had caught word of the tsunamigenic earthquake and had begun evacuations on their own. Electricity was cut, coastal train lines were halted, and military personnel went through villages on foot warning people. While the Disaster Management Centre largely claimed these actions as a success due to their early warning training efforts and information dissemination, one report

FIGURE 2.2. Political cartoon, *Daily Mirror*, April 12, 2012. Used with permission of the artist.

evaluated that "it is doubtful whether the government can claim credit for that awareness." The lack of coordinated response was interesting given that the Disaster Management Centre had just moved into its new "improved" premises in Colombo (Disaster Management Centre 2012).

The cartoon also points to the fact that, as the independent news blog Groundviews.org reported in April 2012, the "horrible rise of disappearances in Sri Lanka continues unabated" (*Groundviews* 2012). In February fourteen abductions had been reported, and in March there were fifteen, adding to a total of 56 disappearances in the previous six months in Sri Lanka. In some of those disappearances, the government's involvement was confirmed. Today, Sri Lanka has one of the world's highest numbers of forced disappearances, between 60,000 and 100,000 since the 1980s, beginning with the violent quashing of the JVP Marxist uprising and continuing through the civil war.

These exchanges of power, the dialectics of threats and doubt, the complex interaction of anticipatory states, all speak to the reach and limits of the government's anticipatory measures. Anticipation is not only evidenced by the ways in which people are prepared: bags and ID cards safely packed with a change of clothes. Security resonates differently when considering how people rely on their neighbors, the sound of the ocean, the gray color of the sky, the tug of memory, the blackness of the rising ocean tide, past warning failures and hasty evacuations from danger—the conditions that shape how people are attuned to the future *and* the past. Sri Lankans know very well the fragility of peace. They anticipate that something else could always be—a condition of living in a place that has been proven sometimes cruel and often difficult. These encounters and exchanges of what I call *anticipatory states* highlight not only the changing structures in which Sri Lankan disaster nationalism manages the risks of tsunamis and terrorism but also how anticipation, as a state of being, a being that has long been a way of life in Sri Lanka, coalesces and pulls back on the anticipatory state. Anticipation is a persistent awareness that also critiques the social conditions that reproduce forms of insecurity people must negotiate. Everyday anticipation wrests the total control and management of the nation away from those dangerous forms of state power that still lurk and have long been a part of life in Sri Lanka.

Najeema put it quite succinctly to me on one of my visits to the village where the first tsunami warning tower was built. We sat together in the small community center built after the tsunami disaster. The warning tower stood next to the center, protected by metal fencing. Playing children screamed and laughed around us, attending the small village Montessori program that she ran here. We were discussing how the warning towers had been working, and as I began to talk about the complex governmental network of experts, scientists, and meteorologists involved, she interjected, "The scientists, the experts, they can know about earthquakes and the tsunami after they happen, but only God knows if they will happen beforehand. And only God knows if the tower will work."[13]

The straight, uncertain path can seek only an endless horizon.

V. V. Ganeshananthan

Rubble Road, July 2005. After the tsunami, pieces of broken homes were used to re-create this road running along the beachside. I was told that some of the palmyra trees had lost their "heads" in the tsunami. Photo by author.

For it is not in the name of a better or truer world that thought captures the intolerable in this world, but, on the contrary, it is because this world is intolerable that it can no longer think a world or think itself. The intolerable is no longer a serious injustice, but the permanent state of a daily banality. Man is not himself in a world other than the one in which he experiences the intolerable and feels himself trapped.... Which, then, is the subtle way out? To believe, not in a different world, but in a link between man and the world, in love or life, to believe in this as in the impossible, the unthinkable, which nonetheless cannot but be thought; "something possible, otherwise I will suffocate."

Gilles Deleuze, *Cinema 2*

3 Endurance

Duration, the time during which something continues, refers to a qualitative multiplicity—continuity and heterogeneity (Bergson [1896] 2004; Deleuze 1992). In this chapter, endurance gestures toward the ongoingness of disasters—tsunami and war—and the experiences of enduring them in Kalmunai. In particular, I highlight how modes of disaster governance and management create enduring forms of insecurity that Sri Lankans living on the east coast must negotiate, explore the practices of living with and within these circumstances, and consider people's efforts to persevere. Here the state's homogeneous time of disaster nationalism renders the future as always potentially disastrous, while simultaneously erasing the sedimented histories of violence perpetrated by state-sponsored militarism and militarization.

I have a shared commitment with other anthropologists who are interested in forms of social life that maintain the force of existing, and in the efforts it takes to endure at the margins of the state (Das and Poole 2004), inside the spaces of exclusive inclusion and alongside the rational technopolitics of nationalism where there might be no transcendent intentions outside of living and surviving (Banerjee 2020; Povinelli 2011; Wool 2017). And in line with anthropologists who suggest this commitment is possible by descending into the ordinary (Allison 2013; Das 2007; Stewart 1996, 2007), I explore, ethnographically, what Beth Povinelli highlights as "forms of suffering and dying, enduring and expiring, that are ordinary, chronic, and cruddy" (2011, 13). But instead of expiring, and dying—which, to be sure, have been all too present, sweeping, and dramatic in Sri Lanka—my focus here is on the rhythms and experiences of living. I examine the crumbling ruins of the tsunami, broken foundations of homes, military checkpoints, hot tin shacks where people live as they await their newly built tsunami homes (Navaro-Yashin 2012). Through these experiences, we can see that state fantasies of disaster nationalism further engender everyday and ongoing conditions of insecurity, uncertainty, and anxiety. As Becky Walker (2013) has explored among war- and tsunami-impacted communities in Batticaloa, I also am interested in the ways in which Sri Lankans *endure* violence. Recognizing the catastrophic and crisis-laden as more than events, I suggest that they *are* the ordinary and chronic (see Berlant 2011; Wool 2017) in the east. Situating disasters as both the presence of the past and the always-possible future, I reveal the distinct knowledge and awareness that have become the infrastructures of enduring and endurance (Choi 2021). That is, this chapter highlights the multiple forms and practices of enduring and endurance.[1]

The chapter's two main sections draw out the temporal and heterogeneous chronopolitics of living in Sri Lanka after the tsunami and during/after the end of the war. First, I discuss what I am calling "slow life" to draw attention to the ways in which life—not just in the biological sense—persists in contexts of insecurity and social and political anxiety. I move through experiences of slow life amid post-tsunami housing and reconstruction practices in the east, after the enactment of the government's no-build buffer zone in Sri Lanka, tracing how endurance encapsulates the syncopated temporalities of disaster, memory, and violence. In the second section, I shift ethnographic form. I present a series of text messages distributed by the Sri Lankan government, coordinated with the Ministry of Defence. These messages illustrate attempts to create a "unified" national consciousness as the Sri Lankan Army moved closer and closer to their victory against the LTTE. I juxtapose these messages with excerpts from daily conversations taken directly from my field notes corresponding to the dates the messages were sent. This section and these juxtapositions more performatively illustrate the chronopolitics of disaster nationalism: enduring forms of (slow) life in a disaster-obsessed state. Acknowledging the banality of disaster is not a failure to recognize what is unacceptable (see Walker 2013). After the horrors of World War II, Deleuze (1989) found that it was the intolerable of *this* world that could lead to new thoughts and new sensibilities (see also Dumit 2014; Rajchman 2007). Inspired by how Deleuze located new sensibilities and affects of time and history in and through cinema, this chapter shows how various forms of endurance cannot be reduced to or determined by conventional linear narratives of history; they are the possibilities of enduring the persistent disasters of state-sponsored nationalism.

SLOW LIFE

In this section, I refer to slow life as the daily adjustments, movements, forms of labor, and memorializing—making live, not (just) in a biological sense—of disaster-affected Sri Lankans. I give particular attention to how east coast Sri Lankan Tamils and Muslims persevere, the ways acts of endurance show how a "sense of life supports an affirmation" (Arif 2016) rather than a denial of life or letting die. Thus, this chapter examines not strictly power *over* life, but rather the power *of* life itself (Deleuze 1997; see also Agamben 1999).[2] The experiences described in this chapter suggest that making a life includes the making of minor adjustments and incremental changes and the creation of regularity and habits (D'Avella 2019; Grosz 2013; Maunaguru 2019) in an environment of perpetual insecurity. These stories gesture to the "practices of making do in a protracted moment of dire and even life-threatening uncertainty that seems so relentless it becomes ordinary" (Wool 2017, 80).

Slow life is also inspired by Lauren Berlant's notion of slow death: the "physical wearing out of a population in a way that points to its deterioration as a defining condition of its experience and historical existence" (2011, 95). This describes the conditions of the poor, laboring, and often racialized or minoritized classes of people who, as they attempt to carve out a space for pragmatic life-making, are simultaneously bound up in the structural conditions of labor and violence that contribute to their slow attrition of life. In marking this tension between life and death, Berlant aims to open up a space that asks us to rethink notions of causality, subjectivity, and life-making beyond or not limited to notions of agency. As such she asks us to consider everyday activity in the spaces of the ordinary that do not follow the logics of effectuality, resistance, or lifelong accumulation. For Berlant, then, survival is slow death, and reproducing life is not necessarily about making it or oneself better. In the context of Sri Lanka, I focus on slow life because death—with over 30,000 dead from the tsunami and 65,000 from the war—has been so overwhelming and overdetermined and has been the incitement to so much intervention, both international and state-based. This is not to say that slow death is not present in Sri Lanka—indeed, years and years of war, for example, have certainly created conditions of poverty, fear, and exhaustion that many individuals and families have to struggle with. However, in these tensions between life and death, in these spaces of crisis ordinariness (Berlant 2011, 101), I focus on the attritional and incremental ways of life-building—the alterlives (Murphy 2017) of disasters—that do not necessarily follow logics of resistance or intentionality.[3]

On a small strip of land, nestled between the government-implemented sixty-five-meter no-build buffer zone and other newly reconstructed tsunami homes, sixteen families awaited the construction of their post-tsunami flats in a government-built temporary housing scheme. The plot was government land, I'm told, formerly a "rubbish heap." The land hosted a makeshift building constructed of corrugated tin and wood. In 2008, when I first met the families living there, this was to be their third and last temporary setting until their new permanent homes were built. Temporary is an odd adjective to describe the plight of these families, who are forever displaced from their original homes, which the government deemed too close to the ocean. Where their old homes once stood on the eastern coast, a sixty-five-meter no-build buffer zone—a territorially defined strip of land along the coastline in which damaged and destroyed property was forbidden to be reconstructed—had been put in effect by the Sri Lankan government. Immediately following the tsunami, which affected nearly two-thirds of Sri Lanka's coastline, then president Chandrika Bandaranaike Kumaratunga hastily put into effect a uniform buffer zone of one hundred meters along the southern coast and two hundred meters along the eastern coast (see Perera 2005). This initial zoning law was made largely out of the desire to take urgent action in the chaos of the aftermath—not based on damage or environmental factors affecting coastal vulnerability—and because the government claimed the public security of the nation was at stake (Hyndman 2007). Kumaratunga alleged that a uniform buffer zone was the fairest way to prevent people from moving back to risk-prone areas.

In the Eastern Province, tsunami housing reconstruction was delayed and became a complex puzzle because of land scarcity and population density (de Silva 2009). The most devastated villages on the east coast were sandwiched between the ocean to the east and the A-4 highway and long stretches of paddy fields to the west. With the implementation of the no-build buffer zone, where would all these people rebuild their homes? Given these struggles and the fact that the uneven demarcations of the buffer zones began to fan the already glowing flames of ethnic tensions (Why only one hundred meters in the predominantly Sinhalese south and two hundred in the predominantly Muslim and Tamil east?), the government began to reconsider how to fairly redraw the no-build buffer zones (Le Billon and Waizenegger 2007; Hyndman 2007). After Mahinda Rajapaksa was elected president in November 2005, the buffer zone "set-back" standards were relaxed. This was in keeping with the extant, although unenforced, regulations created by the Coastal Conservation Department (CCD) in 1997. The buffer zones demarcated along the coastlines by the CCD ranged from thirty-five to 125 meters from the mean high-tide water mark. So along much of the east coast, the buffer zone was reduced to sixty-five meters. Despite this relaxed policy, resettlement and scarcity could not permit a simple moving back of everyone away from the ocean by sixty-five, much less two hundred, meters. Hence, a variety of measures were taken to resettle families and entire villages.

Though the tsunami had been thought of as an opportunity to "build back better," the drawing and then the *making* of buffer zones to build a nation better prepared for future tsunamis also created other forms of social instability and insecurity.[4] In the aftermath of the tsunami and amid the throes of reconstruction, many did not have the privileges of choice, freedom of mobility, and accessibility. In the struggle to give everyone "a place," reconstruction and resettlement also served to cut off, redirect, and stymie movements. In war-affected spaces, where mobility and freedom of movement had already long been issues, disaster management and reconstruction also became part of the crisis.

Some families, enticed by the offer of a new home, agreed to move with their kin and village inland; some communities moved as far as twelve kilometers away from their old homes and away from the buffer zone. Moving inland was difficult in many respects; the need to fill in wet and marshy paddy land extended the already drawn-out process of tsunami home reconstruction. With these moves, livelihoods also had to adjust. Distance from the ocean made fishing a more difficult or no longer viable job option. Some had to turn to agriculture, planting crops in their new neighborhoods. Checking out one newly constructed village eleven kilometers inland, I met a family and discussed their experiences in their new home. The matriarch of the family, Sanjana, told me that she thought about her old home daily: it was the place where she grew up and where she raised her family. She missed the ease of life there. In her old home, it was easy to get her daily tasks accomplished, get to the shops, and go to the doctor if there were illnesses in the family. In her new village, which was far away from town and isolated, it was too expensive to ride the bus to go to the market or visit the doctor. Despite her fears of another tsunami and the sad memories of destruction and death, she told me that she would prefer to live in that old home. This was not allowed, and, in any case, she did not have the means to rebuild it.

In the municipality of Kalmunai, the solution that impacted a group of people I came to know was the construction of "tsunami flats." Kalmunai, sandwiched between the rice paddies of Sri Lanka's agricultural heartland and the ocean, did not have enough land to spread people out, so instead, the government and international donors started to build up. At first, people were opposed to the idea. Rani, a woman in her fifties living with her husband in a temporary housing scheme while waiting for her tsunami flat, would say: "This is not Colombo. This is not the city. We are not accustomed to living this way." But over time, without any other options and in many cases *still* stuck in cramped temporary shelters five years later, a tsunami flat would do.

The heat in the temporary homes was suffocating; the tin walls of the makeshift semipermanent structures absorbed and radiated heat. In heavy rains, the shallow concrete foundations would easily overflow. Each "unit" was one room, at most about twelve feet by twelve feet, although I never took measurements. Families occupied multiple units next to each other. All units shared walls and a roof, and privacy consisted of hearing every neighbor's noises. A large water tap had been set up for all families to use. On one side of the temporary shelters, a makeshift veranda or covered outdoor area had been created. It was here that I spent much time chatting with those living here. It was outdoors, covered and shaded and therefore a bit cooler. From the temporary units, through the palmyra trees, past the broken foundations

FIGURE 3.1. Post-tsunami temporary housing scheme. Photo by author.

of leveled homes and upturned wells you could see the ocean. Next to this makeshift veranda was Selvi's tiny shop.

Selvi, ever the generous shopkeeper and giver of warm Coca-Colas on hot days, told me about how she lost her daughter in the tsunami. That day they were together, and when the water suddenly rushed in, Selvi grabbed onto a palmyra tree and instructed her daughter to hold tightly onto her, but the water was too strong: it swept her daughter away. Selvi herself was injured and spent a month in the hospital. Her daughter's body was never recovered. On the anniversary of the tsunami, Selvi goes to the temple for a special *puja* and then gives food and clothing to the local orphanage. She says it is because so many people lost their children that she and her family give to children on that occasion. The *puja* is also for her daughter. She explained it to us: "When her soul is roaming, sometimes she will feel sad because she feels, 'Oh, mother and father have forgotten me,'" but, after a *puja*, her daughter's soul will feel remembered again. Even in heaven, Selvi told me, the soul can feel sad and sorry.

Some time after Selvi's daughter passed, a fortune-teller walked through the village and, passing by, looked at her and said that there was a soul following her. Selvi understood this soul to be her daughter. It was fitting, she thought, as her daughter had earned the name "little cow" because she was always following her mother around. She told me that now she lives for the future of her surviving elder daughter, maintaining her small shop selling individual packets of laundry detergent, shampoo, phone cards, biscuits, and bottles of Coke and Fanta. If there is another tsunami, she told me, she will run. Again.

The past literally follows her, a reminder of something, someone, that she will never ever forget, even as she prepares to move into a new home away from the shoreline. And in Selvi's story, we can tease apart the interwoven temporalities of life, disaster, and disaster nationalism in post-tsunami and postwar Sri Lanka. There is the past that will always stay with her, the possibilities for the future, the difficulties of the present, the potential for another disaster—the temporalities of disaster and life.

On the other side of Selvi's shop, smoke bellows from a shallow black pan that is being shaken by deft, wrinkled hands, taking care that the roasting peanuts do not burn. Under the veranda that provides cover from rain or shade from the searing tropical sun, Sitamma crouches as she lets the roasted peanuts cool and bags them to later sell to hungry schoolchildren. Usha (my research assistant) and I try not to disturb too much, lest we interrupt Sitamma's counting, for she must keep track of how many peanuts go into each little plastic bag. "It is nothing much," says Sitamma. She sells the peanuts for a few rupees—a little bit of money to go toward basics such as food. Mainly, she says it keeps her busy and is something to do. Indeed, in the mornings and afternoons when we talked, I always wondered if I was bothering her, but she insisted that I was not. Over the course of many visits and conversations under that awning, Sitamma described to me how she was taken up by the waves near her home, but somehow was carried over to the courthouse, near the police station. She was slightly injured, and she did not get sick from swallowing dirty tsunami water like many others did.[5] Sitamma recounted that she woke up in the hospital after the tsunami. Drawing her hands up to her ears, she told me that she survived with just her earrings on. Even though she said she never goes back, she recalled her previous home, with its high brick walls and privacy, its space, and its luscious garden, fondly.

The Eastern Province of Sri Lanka, in which this housing scheme was situated, was always a specific geographical fixture in the war (Fonseka and Raheem 2010; Hasbullah and Korf 2013; McGilvray 2008). It was, as I ex-

FIGURE 3.2. Sitamma preparing curry leaves. Photo by author.

plained before, considered part of the Tigers' imagined homeland of Tamil Eelam—a problematic notion not least because the eastern region, especially the coastline, hosts a significant Tamil-speaking Muslim population. And though the Eastern Province was finally "freed" from the "clutches" of the LTTE by the government in 2007, by 2009, following the end of the war, signs of ongoing securitization remained visible. With the state of emergency still in place, with checkpoints, roadblocks, and the Special Task Force (STF) still ever-present, tension was still visible and palpable. As discussed in chapter 2, the possibility of "terrorists lurking" justified the ongoing presence of the military and the surveillance of Tamils, especially young Tamil men.

The 1990s were a particularly violent period between the LTTE and the Sri Lankan government, characterized by brutal retaliations between the two parties in the east. And in particular, in Kalmunai, where this post-tsunami housing scheme was built, many civilians bore the brunt of these tragic circumstances: the rounding up and disappearances of young Tamil men, burned-out shops. A former *grama niladhari* ("village officer") recalled that it was too difficult to identify bodies found at that time; so many were unrecognizable. The ocean, he remembered, was tinged red with blood and littered with headless corpses. Though the security situation in 2008 and 2009—the time period in which the fieldwork for my research was conducted—was certainly more relaxed, and has improved even since then, for many Tamils, memories of these "hard times" sometimes made their way into the present. The '90s *prashanai* (problems) were also recalled by Sitamma under the makeshift awning. She told me about the last time she saw her two sons, who were taken by the Sri Lankan Army in a "roundup" of young Tamil men. After months of almost daily visits, I had learned that she lost two sons "in the war" but I had not pressed for details, until one day the story, like her tears, poured out. It was the first and only time I saw Sitamma cry.

She recounted that during those years, many lost sons and daughters. Some families lost two or three children, "nearly like the tsunami deaths." It was July 7, she said, and the Sri Lankan Army had been going through villages and rounding up all the young Tamil boys in the nearby cemetery. The boys were then all put into a truck. When family members stopped by, the army would chase them away. "Maybe my sons would have seen me," Sitamma wondered, "but I couldn't see them." The army claimed that they didn't have her sons when she asked them where they were. When the truck started to move, Sitamma said she ran after it but was obstructed by the police: "*Odi, odi, odi!*" (Run, run, run!), they shouted at her. She never saw her two sons again. Sitamma tried to find them; for two or three years she said she looked for them. She collected the proper documents and files and sought the assistance of the police. She visited offices in Colombo, the nearby army base Karaitivu, and the capital of the Eastern Province, Ampara, asking about her sons. Everywhere she went they all said, "No, no, we don't know of these boys." She was finally given monetary compensation in Colombo for the loss of her sons: thirty thousand Sri Lankan rupees, which amounts to approximately three hundred US dollars. She said that they would have been thirty-five and thirty-two years old.

Sitamma thinks of her sons all the time, she says, but she does not always share her feelings or thoughts. At family gatherings, she especially wishes that her sons were with her. She knows she is not the only one who suffers: "Everyone has this sadness. Everyone has feelings of missing their loved ones." People go on with their daily work, Sitamma claims, because without work, they will be too sad. And so she says if ever she has sad thoughts, she tries to divert them by keeping busy. Indeed, in the mornings we would often chat while she did her work: roasting peanuts and bagging them to sell to schoolchildren for five rupees; preparing the ingredients to make curry powder; preparing curry for lunch and dinner; sweeping her tiny one-room shelter; watering her small garden.[6]

When the war ended, Sitamma shrugged her shoulders at me and said, "What has changed?" Sitamma always insisted that if another tsunami came, she would, pointing to the ground, "stay right here." She explained, "I have lived through so many things—a story that is too big to tell. I lived through a war, a cyclone, a tsunami. My daughters are married, my sons and husband are dead." She told me she wasn't afraid of anything anymore, that the only thing she had not experienced was her own death, and that she would pass the time until it came by doing her daily activities and patiently waiting. She would wait on her terms.

The destruction and toll wrought by the tsunami and war were expansive. Official numbers proliferate in a litany of death figures: over thirty thousand dead after the tsunami, over sixty thousand casualties of war. Scaled down to families, the numbers are still staggering. In the east, the tsunami claimed fourteen from Fareed's family, five from Selvi, two for Tharanga, eight for Ravi. In the more distant past, a number of the Tamil families I came to know had lost many of their male kin. Most of them had disappeared, likely taken by the Sri Lankan Army: two sons, a brother, an uncle. The devastation adds up, on a universal and intimate scale. But to say that the everyday is bogged down by these demons paints a pallid image of their everyday. Upon these memories, lives continue to thrive and persist. They too are forgotten, only to make themselves remembered when the occasion calls for it (see Simpson 2020).

Mohan was living in his sister's family's home across the lane from the temporary housing scheme. His old home, or what was left of it, was in the buffer zone. He took me to the site where he used to run a small shop before the tsunami leveled it. Looking at the crumbling foundations that remained, Mohan pointed out that it was now "just grass." Mohan lost his wife and two children in the tsunami and because of the buffer zone was unable to rebuild his home. He now stayed with his sister's family, trying to earn a living. At the time, in 2009, looking for work was difficult: "If we want to live, we should live in a happy and peaceful way. Now if we want to travel, I have to get down because of checkpoints, and there are always problems." He explained that because he could not travel freely, it was difficult even to fish or to attempt to get any other job: "We are just staying here. . . . How can we say the war is stopped? If the war is stopped, we should be able to travel freely for our needs, but with the checkpoints we must wait so long in the hot sun."

Mohan was commenting on the continued presence and power of the STF in the east. In addition to conducting neighborhood and house checks in search of hidden terrorists, they also maintained roadblocks and checkpoints, obvious markers of surveillance that, like the tsunami warning towers, dotted the physical landscape of the island. Checkpoints made traveling extremely difficult for some, illustrating both temporal and spatial dimensions of disaster nationalism. For example, leaving the east to go to the capital, Colombo, could take up to twelve hours on the bus. The same trip reversed took about eight hours. This was because it was assumed that those making the journey out were mostly Tamil and thus marked by suspicion (Thiranagama 2011; see also Jeganathan 2000). The roads leading out of the east were monitored by the Sri Lankan Army. In the first twenty kilometers, the bus would stop at least five times at checkpoints. At each checkpoint, everyone, unless elderly, a mother traveling with a child, or a foreigner like me, had to get off the bus. Bodies were frisked and bags opened and inspected. The bus was also examined. During the day, the heat was relentless, stifling whether you had to wait in the hot bus or line up to wait for your inspection without the respite of shade. Many made the journey at night, when it was cooler, although nighttime hosted its own set of safety concerns. Waiting functions as a form of state and military power, stealing time and imposing barriers to and delays in everyday life.[7] Checkpoints figure as both physical and imagined barriers for travelers. Tamil women, especially, were weary of travel; they had heard tales of violation and physical assault by military men at checkpoints (Hyndman and de Alwis 2004).

Mohan did not have the funds to start up the shop he had run out of his home in a new place. He described his situation and change in circumstances, a description that was relayed to me by many others: "People who were poor are now rich and rich people now poor." He was given a loan from an aid organization, but it was not enough, so he returned it. Instead, Mohan started going with other men on their boats to fish. As someone who did not own a boat and had limited fishing experience, Mohan likely was not getting paid much. Given his lack of experience at sea and what he had lost in the tsunami, I asked him if he feared the sea. He responded: "Why fear the sea? Going to sea is like going to war. If you come back, it's a victory. If not, that

FIGURE 3.3. Mohan leading me to his old home by the sea, now "just grass." Photo by author.

means death. We have fear living here, because of war. But under the roof of the sky we have made our home here and we are living and adjusting."

As the grass slowly overtook the remnants of Mohan's home, President Mahinda Rajapaksa expressed how keen he was to move on quickly and bury the rubble of the past. Two months after the end of the war, in an interview with *Time* magazine entitled "The Man Who Tamed the Tamil Tigers," Rajapaksa's response to a query regarding a possible war crimes investigation in Sri Lanka was: "I don't want to dig up the past and open up this wound" (Thottam 2009). Meanwhile, Sri Lankans like Mohan would continue to live and adjust.

JUXTAPOSING DISASTERS

As described earlier, this section makes a decided shift in format and narrative register, though it still attends to the chronopolitics of disaster nationalism. Here slow life also plays out, through anecdotes of everyday life juxtaposed with the Sri Lankan government's expediently distributed text messages to the nation, providing succinct, real-time announcements portending their war victory. Technologies of disaster nationalism, as illustrated by the national tsunami warning system in chapter 2, work to draw Sri Lankan citizens into a collective awareness and orientation toward not just disasters but also the nation's war against terror. The consistent messaging attempted to draw the nation into a collective awareness and consciousness about what was happening in the war-torn north, from which everyone, including independent journalists and humanitarian organizations, was banned (more on this in chapter 4). Through this messaging, the government not only foreshadowed the war's end but also aimed to forge a new national imagination, a fantasy of peace and unification, through violent yet "humanitarian" means. I weave these messages with the (slow) lives within this national space-time. These narratives are taken directly from my field notes, from conversations with women and men living in the eastern coastal regions of the island.[8] They are stories collected from both Muslims and Tamils, living in different tsunami- and war-affected villages. I use their voices to pluralize the government's singular nationalist narrative. These narrative juxtapositions can be seen as counterpoints, commentaries, and experiences in shared linear time, but not as a singular history.[9] Here too we see the multiple tenses of enduring disasters.

My mobile phone buzzes. I have received a text message from the Sri Lankan government's Information Department:

> 5.16.09, 4:49 p.m.
> I will return to Sri Lanka as a leader of a country that vanquished terrorism, said President Rajapaksa at the G-11.

Before the civil war ended, Sri Lankan president Mahinda Rajapaksa was out of the country attending the G-11 meetings in Jordan, proudly proclaiming the imminent success of his militaristic efforts to finally rid Sri Lanka of the LTTE "terrorist" problem. In fact, throughout April and May much of the mobile-phone-using population and I had been receiving updates from the government on their progress purging the LTTE from their last territorial stronghold in the north.[10] On the news and on their website, the Ministry of Defence boldly broadcasted that this last month was "The Final Countdown." The areas of the north where the government was routing out the LTTE were completely off limits, not just to other civilians but also to independent journalists and humanitarian organizations. Even the International Committee of the Red Cross pulled out of the area after a hospital was shelled, citing that it was too dangerous for their workers to deliver aid to those in need. Those in particular need were wounded civilians, some of whom were being used as human shields by the LTTE. The LTTE and humanitarian groups claimed that civilians were being shelled by the Sri Lanka Army. Accusations were denied, and people could only speculate about what was really happening in the war zone. In a savvy employment of technology, the Sri Lankan government endeavored to create a synchronized national consciousness through the distribution of text messages and updates. Though the rest of the nation was not actually in the north, text messages connected people to places where they were not and to people they did not know, cultivating a collective anticipation of the end of the war. Perhaps everyone in Sri Lanka could be coordinated in the homogeneous time of the new, united, and peaceful nation.

In these messages the government took great efforts to publicize their "humanitarian mission" of saving Tamil civilians and made claims about what was *really* happening in the war zone.

> 5.1.09, 5:25 p.m.
> President urges civilians to enter Gov't controlled areas, in a leaflet air dropped to the No Fire Zone

The country was privy to the imminent end of the war. Over the course of several months, the gradual overtaking of the Tigers and the shrinking territory under their control was publicized. On the Sri Lankan Defence website, the army boasted about undertaking the "World's Largest Hostage Mission." The last two weeks of messaging detailed the Sri Lankan government's aggressive and tactical efforts to rid the island of terrorism, once and for all.

> 5.10.09, 5:47 p.m.
> LTTE attacks using heavy weapons in the No Fire Zone have caused immense damage to the civilians entrapped in the area
> –Military–

The numbers of civilians they were saving grew as the army closed in on Tiger territory and sequestered civilians into a designated "No Fire Zone" (NFZ) or "Civilian Safe Zone" (CSZ).

> **5.8.09, 6:32 p.m.**
> Previously defined "Safe Zone for Civilians" redemarcated to match current realities: new area, 2km by 1.5km
> –Military Spokesman–

May 11, 2009:
"The government is only interested in war," says Salowdeen. His friend chimes in, "After 5 years, Sri Lanka will be like Somalia or Bosnia." They say they can't expect the war to be over: "war is like a cancer, it will spread. Stopping the violence is difficult, and peace won't come by war," Salowdeen says. His own suggestion was that if the government had simply been forthcoming about what they would actually do as far as reparations and rights for everyone and every group were concerned, things would be different. But as it is, he said that at the peace talks, one group was "talking about the head, while the other was talking about the leg," and during this time the war never actually stopped. It is all "just talking."

When I asked what he thought at least about the liberation of the Eastern Province, Salowdeen looked me narrowly in the eyes and asked, "Who told you the east was liberated?"

The government kept the public apprised of the army's progress, detailing troop movements and the shrinking of LTTE-controlled areas via these text messages.

> **5.14.09, 9:57 p.m.**
> All civilians held hostage by the LTTE will be liberated within the next 48 hours–President tells Sri Lankan community in Jordan

May 14, 2009:
Priya tells me, security-wise, things are "normal"—in the sense that the STF comes around regularly. Just last Monday (today is Thursday) the STF came and

did their routine ID checks. They come around to make sure no one is "new" to the area. Each house has a form given to them by the G.S. [grama sevaka][11] that lists all their household members. So the STF comes around and looks for that form, and checks to see that there is no one suspicious or not registered in a household. If there is a new person or someone from another area, then they are often taken for questioning. If someone comes for a visit, permission must be granted by the police, otherwise there might be too much trouble.

When I asked her if she had any fears about the STF (Special Task Force), at first she said that she did not, but admitted that sometimes when she first sees them, she feels a little nervous. When the STF comes, all the ladies go outside of their homes and wait. While they wait outside, sometimes she gets a little afraid. She has heard stories of police and STF "behaving badly" with women.

Priya also offered us information she had heard regarding the treatment of women in the north. "People are talking" and saying that women who are crossing the border into Government-controlled land are getting injected and knocked unconscious, and then "bad things" are done to them. When I pressed about which "bad things" were being done, she said that women were getting injections to make them sterile. "They want to stop increasing the Jaffna generation totally." When I asked where they were getting this information, Priya said that some others had seen this on the internet. She also mentioned that 3–4 days previously she had been told that the Red Cross, while giving food and keeping children safe, had been bombed by the Sri Lankan Army, and that even children were killed. She says she doesn't watch TV, but gets her news by chatting with other people. She doesn't know what will happen here in the east, but she says she gets sad when thinking about the people up north: "They are also people; people who also want to eat and be happy."

5.15.09, 11:12 a.m.
Rescue mission near completion. 4500 civilians escape LTTE grip, arrive in Govt. controlled areas during the past 48 hours

May 15, 2009:
Fajiana and her sisters live in another set of flats in the predominantly Muslim area of Maruthamunai. Living in the flats is not easy, they say, but what can they do? They moved into the flats about a year and a half ago. Before

then they were in temporary shelters, just on the other side of the school, in the village: "We don't have any other place to live. We don't have any land to build on. Obviously it is not as comfortable as a single house, but still, it is a place to live." The flow of water is inconsistent and the toilet inside feels "like a teabox." Living so closely together, there are issues of noise. Children do not have space to play. However, as Fajiana and Ruzfina tell me, "we agreed to live here, so we have to adjust."

> 5.16.09, 12:52 p.m.
> Large number of civilian influx to Govt. controlled areas, recording over 16,000 during past two & half days

> 5.17.09, 8:09 a.m.
> Civilian influx into Government controlled areas recorded as over 50,000 upto [sic] now since 14th of May

> 5.17.09, 2:17 p.m.
> Latest update: over 72,000 civilians rescued since Thursday. Remaining LTTE terrorists confined to less than 800 Sq meters

> 5.18.09, 6:49 a.m.
> All civilians rescued, LTTE confined to 300 square meters

With the demise of the LTTE seemingly imminent, the days leading up to the declaration of the end of the war also built up an imagination of a Sri Lankan nation, purged of "terror" and completely under the control of government forces.

> 5.18.09, 2:20 p.m.
> All heads of Security Forces to formally inform President Rajapaksa of the historic military victory this afternoon

5.18.09, 3:13 p.m.
Govt requests all state institutions to hoist the national flag for a period of one week to celebrate the victory over terrorism.

The Sri Lankan government declared May 19 a national holiday. Sporadic rainfall and clouds dampened what was supposed to be a celebratory day. Other parts of the country had erupted in joyous celebration. Some people mentioned that in the Muslim neighborhood just a few kilometers south of where I stayed, the Sri Lankan Army had orchestrated a war victory rally. While I could not verify if the rally had been ordered by the army or the STF, I managed to pass by the area on a scooter. From the main road, I could see that empty buildings and quiet storefronts were decorated with posters of Mahinda Rajapaksa's face, but I was too scared to take a photo.

I made my usual round of visits that day. Lakshmi, six months pregnant, shared in chapter 2 that cloudy days made her nervous. Today, also cloudy and gray, was making her feel a little nervous, although she admitted that these last few days, everyone had been feeling tense. It doesn't feel like the war is over, she tells me. In her twenty years of living, she has only ever known war: "The war has been continuing for so long that it will continue to go on. During the war and after, one person experiences happiness and one person experiences sorrow; it will flip flop like this between sides." Sitamma shrugged her shoulders and asked me to look around: "What has changed?" "Anyways," she continued, "we cannot tell anything about the future. We can only tell about what happened in the past."

END(UR)INGS

> The concept of progress should be grounded in the idea of catastrophe. That things "just keep going" is catastrophe. Not something that is impending at any particular time ahead, but something that is always given. . . . Hell is not something that lies ahead of us, but this very life, here and now.
>
> —Walter Benjamin, *The Arcades Project*

Though Benjamin was of course not literally referring to natural disasters, I use this notion of "catastrophes" to point to the ways in which disasters

FIGURE 3.4. Broken clock tower, eastern Sri Lanka. Photo by author.

continue to live on in Sri Lanka, their effects compounded by the projects of disaster nationalism, projects facing a future horizon that buries the past such that disasters can "just keep going."

In the middle of a roundabout in a town on the east coast of Sri Lanka sits a defunct clock tower, its clock face shattered (see figure 3.4). In the late 1980s, in an effort to open up Sri Lanka's economy to foreign investment and in line with IMF structural adjustment policies following the country's economic liberalization, then president Premadasa put in effect the 200 Garment Factory Program (200 GFP), which sought to open garment factories in rural areas of the island. In addition to new infrastructures of paved roads and electricity, and, most importantly jobs, clock towers usually arrived with these factories. Rumored to have been "obsessed" with time, Premadasa wanted the clocks erected to evoke a sense of modern, disciplinary, industrial time, a signal that Sri Lanka too would be participating in the global economy (Lynch 2007). The clock towers, I offer, can be seen as a technology that would attempt to synchronize Sri Lankans into the linear time of capital and the linear homogeneous time of nationalism (Anderson 1991). However, whatever promises economic liberalization held for the country in the 1980s would be quickly overshadowed by other virulent forms of nationalism, with the onset of the civil war in Sri Lanka by 1983. By 2006, eighty-nine of the country's 152 clock towers were in disrepair, and the country was still mired in war.

I do not know anything definitively about this clock tower. I do not know how its face was broken and shattered, the ordering practices of the nation paused at 6:50. While I think the impulse to render this frozen calendrical time as the rejoinder to the passing of linear homogeneous time (Benjamin [1968] 1986) is appropriate, I see in this shot-through, broken clock face still other possibilities: the dreamed-of time of national progress; the time of revolution; and most significantly, the fragile, fraught, and fragmented national history of Sri Lanka. The marred clock is a vestige of institutionalized attempts toward a fantasy of Sri Lankan nationalism. The broken clock, then, also points to the enduring and ongoing crisis of nationalism in Sri Lanka.[12]

Despite President Rajapaksa's call to forget the past and move on with a clear national conscience, people continue to suffer from the ongoing effects of both the tsunami and the war. While these effects were especially acute during the end and the months following the end of the war, many today still await the justice or even the prosperity promised to them by the government. I have juxtaposed the experiences after the tsunami and war to elucidate how programs and policies implemented by the Sri Lankan state to safeguard people and territories from both natural disaster and so-called terrorism have instead created more insecurity. Yet this insecurity does not always manifest in broad strokes of violence or irruptive moments of danger; it often looks more like standing outside in the hot sun while bags and homes are searched. It is making do with a temporary shelter, waiting for a new home nearly five years after the tsunami. It is sometimes remembering, sometimes forgetting the disasters and also knowing that disasters are always possible. In Sri Lanka, disasters, both natural and human-made, live on, highlighting the ways in which disasters are entwined into history and politics, spiraling and enduring in manifold articulations and directions. Disaster, thus, is the event and nonevent, spectacular *and* banal, punctuated *and* chronic. Disasters are rendered in multiple pasts, presents, and futures. In Sri Lanka, disaster nationalism is the imminence of disasters as threats to the nation *and* the ongoing disaster that is nationalism.

The national time of disaster in Sri Lanka is the preemptive approach of disaster management and preparedness; it is the linear modality through which the imagined community of the nation is conjured. From this future orientation and the conditions of disaster and war in Sri Lanka, persistent threat has emerged, necessitating its continual management, which further engenders the power of the state in the guise of national security, such that Sri Lankans retort: What has changed? I am interested in the "enduring" because there are no endings of disaster in Sri Lanka. And yet the timeless imagination of disaster nationalism is not the only enduring form which determines life in Sri Lanka—indeed, to believe so would be to miss the ways life unfolds in parts of Sri Lanka, however slowly. Living and life choices are modalities of lateral agency (Berlant 2011), effectuating neither transcendence nor resistance but rather a mode of enduring, a "getting by" (Allen 2008) when hardship and insecurity characterize the social conditions under which life must, nevertheless, go on. These experiences are the forms of endurance it takes to live in a hot tin shack, to wait at a checkpoint, to see the grass slowly overtake the foundation of your old home and your old life. For, to return to Mohan, "under the roof of the sky" there is living, there is adjusting, there is waiting. This is an endurance perhaps not of a transcendental or ostentatious kind of vitality, but of vitality nevertheless.

The kite pretends to know a direction, but even those who send it into the sky cannot see the wind of its end.

V. V. Ganeshananthan

Tsunami evacuation route sign, Kalmunai. Photo by author.

> The spectacle is not the collection of images; rather it is a social relationship between people that is mediated by the image.
>
> Guy Debord, *The Society of the Spectacle*

4 Reiteration

In the second century, Ptolemy, the so-called Father of Modern Geography, constructed a map of Sri Lanka—then referred to as Taprobane—which depicted the island upside-down and nearly continental in size (see figure 4.1). Its boundaries, topography, and shape were based upon the collected accounts of seafarers. No doubt the gaze that was turned to this eastern isle conjured up a fabulous land of mystique and spices—an exotic locale to which, years later, the Portuguese, Dutch, and British would be drawn and which they would eventually colonize. Ancient maps of Sri Lanka represented what was known by outsiders about the small island, the "Pearl of the Indian Ocean"—the memories and experiences of worldly explorers in pursuit of treasures. Perhaps even its exaggerated size indicated its allure of highly desirable spices and trade goods, ultimately, a vision of conquest, exploitation, and colonization.

The Dutch, who were first enticed by the island in the early sixteenth century in search of cinnamon and wealth, wrested control of territories along the Sri Lankan coast. The Dutch built forts, introduced missionaries, and, though unable to penetrate the interior Kandyan kingdom, managed to

FIGURE 4.1. "Taprobane," Sri Lanka, second century, as depicted in Jacopo d'Angelo's translation of Ptolemy's *Geographia*. Source: Library of Congress Online Catalog.

force the Portuguese out from Ceylon in 1658 and assume control of coastal Ceylon (Abeydeera 1993). The Dutch East India Company (VOC) immediately dispatched surveyors to map Ceylon's coastal regions, territory crucial to their economic and military interests. The VOC was central to the expansion of the Dutch empire, and concomitant with this expansion was the "golden age" of Dutch cartography (Barrow 2008). Despite their maps' geographically inaccurate representations of the island's shape, the Dutch administration

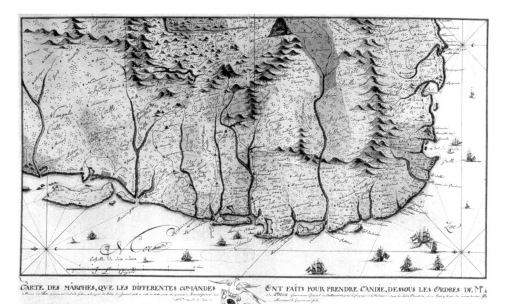

FIGURE 4.2. "Carte des marches que les diferentes commandes ont faits pour prendre Candie, dessous les orders de Mr Le Baron van Ek, gouveneur de Lisle de Celo, et levez par les Ordres le 2 janvier 1765." Map indicating the itinerary of Dutch troops, 1765. Source: Abeydeera 1993.

introduced a new era and style of mapping, indeed a new conceptualization of the island, and "greatly improved upon the Portuguese knowledge of Sri Lanka" (Abeydeera 1993, 101). Dutch surveyors drew views and plans of strategic economic and military importance. Maps detailed, for example, cinnamon plantations, factories, coastal fortifications, and strategic and defensive positions to ensure their colonial power on the island as shown in a map of a Dutch military expedition in figure 4.2 (Abeydeera 1993).

The British Empire, benefiting from the French invasion of the Netherlands in 1794–95, met very little resistance from the Dutch and came to occupy Dutch Ceylon by 1796.[1] While the British did establish a survey department—in fact, by the mid-1850s it was the largest department within the colonial administration—a map of Ceylon was not produced until 1903 (Barrow 2008). The British, however, did adopt a form of colonial governmentality with the Colebrooke-Cameron Commission Reforms of 1832–33. While these reforms could be perceived as a democratization of power, they reflect, rather, following David Scott (2004), transformations of power; the Colebrooke-Cameron Commission Reforms changed the administrative organization of the island by dividing the country into provinces with a centralized judicial and administrative system and dismantling government monopolies. The general thrust of these reforms was to make the colony more profitable. As administrative lines were drawn, allying Sinhalese regions and separating them from Tamil and Muslim regions, notions of race and nation also began to emerge. In part, this was because in the creation and installation of the colonial Colebrooke-Cameron Reforms, religion and caste, which had been the primary markers of social difference, were not recognized. Around the same time, a more "accurate" English translation of the *Mahavamsa*, a mytho-historical text written by a Buddhist monk in the fifth or sixth century, became standardized by scholars. This colonial text-based interpretation cultivated the *Mahavamsa* as a historical text chronicling Buddhist kingship in Sri Lanka, becoming significant for Sinhala nationalists. The text gave authority to a history that posited the Sinhalese as the original inhabitants and Tamils as the foreign invaders (Rambukwella 2018; Rogers 2004)—the mythic history Rajapaksa praised in his war victory speech in chapter 1.

Tracing the cartographic reiterations of the Sri Lankan "geo-body" (Thongchai 1994) through this abbreviated (and far from comprehensive) history offers a way to understand the emergence of nascent national formations—and, in the case of Sri Lanka, of an island nation (see Sivasundaram 2013)—through the territorial governance of colonial states (see Anderson 1991; Krishna 1994; Ramaswamy 2010).[2] As critical cartographic theories and practices have illustrated, the modes of power present in maps and in their productions reflect values and particular historical forms of social order (Harley 1988; Wood and Fels 1992). Yet as postcolonial scholarship has shown, these technologies of power were never absolute or complete; neither were they straightforward or smooth processes of knowledge production and governance (Craib 2004; Edney 1997; Mitchell 2002; Turnbull 2000).

These cartographic imaginations also help us understand contemporary territorial representations of Sri Lanka and their uses and significance today. In these colonial geographic visualizations, we can find particular imaginations and fantasies of Sri Lanka's island territory. The map created in Ptolemy's time signifies Sri Lanka's imagined place and space in the world because it was a focal point of trade, a place that would elicit the adventurous wiles of characters such as Sinbad and Marco Polo and, of course, the colonial forces of the Portuguese,

Dutch, and British in search of spices and control and conquest. While today's digitized maps and real-time image productions are "realer" and seemingly more objective, they also garner the attention of international communities and offer their own modes of engagement and practice and new negotiations of power. So while the technology differs, I suggest that both these forms of maps can be thought of as "fantastic" artifacts and value-laden images (Bier 2017; Harley 2001) worthy of social and ethnographic attention, whose powers both normalize reality and take on unanticipated lives of their own (Mitchell 2004).

These maps also help us to understand the technologies and temporalities of disaster nationalism. In this chapter, I focus on the production of the visual cultures of the tsunami and the war in Sri Lanka: digital maps, satellite maps, grainy cell phone videos. I examine them together as a way of learning the nation (Ramaswamy 2010), of how iterations of disaster maps reify visions of an undivided Sri Lankan nation made continuously insecure by natural disaster and terrorism. First, I discuss the production and proliferation of GIS (Geographic Information System) maps, which were instrumental in illustrating the destruction of the tsunami for use in post-tsunami response and reconstruction efforts. These maps and images were lauded for their ability to show destruction without sensationalism. As immutable mobiles (Latour 1986), these maps were decidedly apolitical, seemingly objective, presuming the geographic stability of the Sri Lankan nation. Here, nation and nature both were uncontested. Just as spices once mobilized colonial powers, these maps drew international interest and attention in the form of humanitarian assistance. As Gennifer Weisenfeld writes about Japan's great earthquake of 1923: "Disaster is...an unfolding temporal landscape within which visual production must be historicized" (Weisenfeld 2015, 4). I then shift to maps and other images produced during and after the war. Unlike representations of the tsunami, these images were highly controversial and heatedly contested. My goal is not to reveal the "real" contents and meanings of these images of disaster, but rather to focus on the social and political dynamics articulated at the level of the visual (Witjes and Olbrich 2017; see also Spyer and Steedly 2013). I do not seek to undermine notions of objectivity, which, as scholars have shown, does not equal truth (Daston and Galison 2007). Looking at the "promiscuous commingling" (Weisenfeld 2015) of the subjective and objective, of fact and fiction and fantasy, I argue that these visual artifacts become more than overdetermined representations of history and culture, but rather are the sites through which Sri Lankan geo-body-politics are repurposed and articulated (see also Eder and Klonk 2017). Following these imaginations of disasters (Sontag 1965) in Sri Lanka is a way to trace how technologies of disaster nationalism work in and through time: from the Sri Lankan nation under threat to the ways that disasters live on, through the state's attempts to manage and control them.

My approach to understanding visual cultures and their politics stems from critical analyses of maps as a technology of state power and as a powerful symbol of nationalism. As value-laden images, maps, like the objects and subjects represented in them, become active agents. Certainly as tools of governance and management for the state, mapping and surveillance technologies must be viewed critically.[3] As sociotechnical objects, maps can perform the "god trick" or "omniscient detachment," and as abstractions are limited by their failure to understand and portray ground reality (Scott 1998).[4] While it may be true that maps are limited in their ability to "see," I am interested in the ways in which maps not only represent but also create realities. Following Thongchai (1994), the map is a metasign, producing meaning and power about the geo-body of the nation beyond a representation of territory. The powerful effects of a map can come both from its "objective" representation of space and from its social and political effects—its reiterations. As a social and political production of nationality of itself (Krishna 1994), then, the map is a visual text; even though it is created in one moment, it takes on its own history. As I will show, maps are constituted by an ensemble of political relations, through which new political possibilities and new conditions of visibility can emerge (Azoulay 2008; Pinney 2015). These reiterations reveal the limits and contradictions of technologies of disaster nationalism. By "ending" the war, disaster nationalism would appear a success, but control of the state's image of peace and security would prove elusive.[5]

MAPPING NATURE, NATURALIZING THE NATION

> It is through the reflexive production and circulation of images that "imagined" social entities like nations become visible and graspable.
>
> —Karen Strassler, *Refracted Visions*

In the early stages of developing this project and during short visits to Sri Lanka immediately following the tsunami, I noticed that mapping, and specifically GIS maps, were ubiquitous.[6] A geographic information system is a system that creates, analyzes, manages, and maps many different kinds of data. A GIS creates maps, connecting spatial locations with specific types of information. New or existing information about a place or space may be added to a database, analyzed, and finally visualized through maps (ESRI 2024). During my visits to different organizations engaged in post-tsunami reconstruction projects, the importance of the knowledge conveyed by these maps was reiterated time and again. Indeed, several agencies came to the assistance of Sri Lanka to help in the production of tsunami maps: the British organization MapAction, the International Water Management Institute (IWMI), and the UN Office for the Coordination of Humanitarian Affairs (UNOCHA), which also created the UN Humanitarian Information Centre (UNHIC), for example. Recognized for its ability to visualize and represent disaster, GIS can not only map spaces of disasters but also align other relevant features, conditions, and events with those geographical spaces. In Sri Lanka, GIS maps were created to show washed-out roads and bridges, to identify usable routes by which to deliver aid resources, and to show locations of existing Internally Displaced Persons (IDPs) camps and even areas with landmines. In disaster management efforts, GIS technology played an important role in post-tsunami activities such as mapping out land-use policies and community resettlement projects and coordinating the presence of humanitarian aid and the geographic locations of their projects. The indispensability of GIS technology in disaster contexts has spurred global initiatives on disaster reduction and recovery. This was especially the case in Sri Lanka, where government officials, university professors, and international aid and humanitarian organizations convened at meetings and workshops to assess the values and progress of GIS and digital mapping technologies in disaster management.[7]

For government and humanitarian agencies alike, these tsunami maps were readily accepted and highly regarded for their utility in the implementation of aid and assistance. And they were continually updated, reflecting changes in the geographic areas they represented: the numbers of IDPs, houses built, camps, injured, and dead. The three GIS-produced maps in figures 4.3–4.5 were created in the tsunami's aftermath and provided "on the ground" assessments.

The first map, created on December 30, 2004, shows the number of people "affected" and the location of agencies; the second, dated December 31, shows the number of deaths by district; the third, dated January 13, again shows "affected persons," but this time with more comprehensive information including

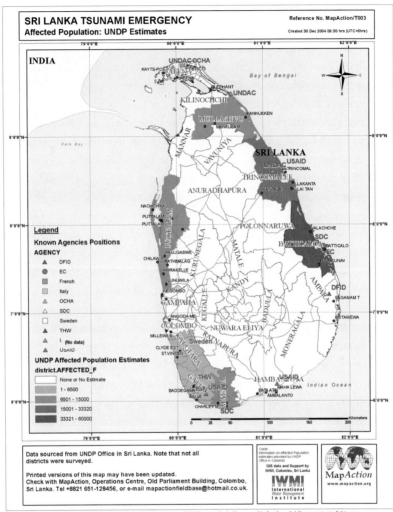

FIGURES 4.3–4.5. Tsunami maps. Source: MapAction.

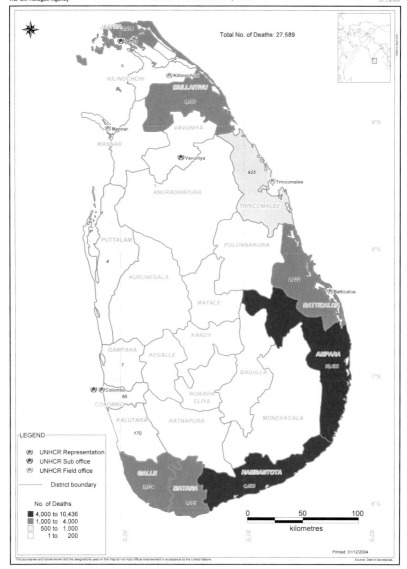

FIGURE 4.4.

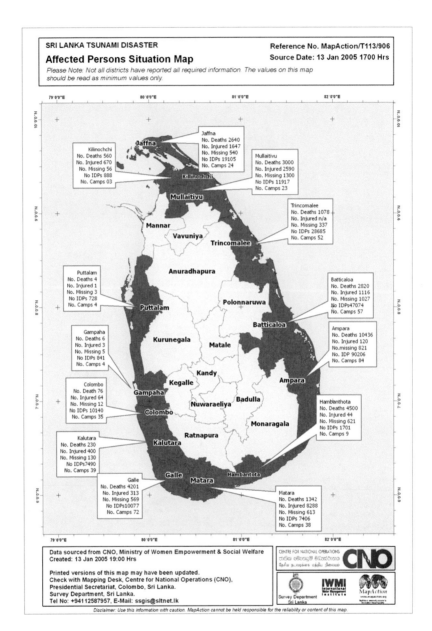

FIGURE 4.5.

deaths, missing, and injured. In these maps, the base map of the island and district boundaries is unchanged; the same map of the island is used to convey different types of information. The representations seem noncontentious and objective, the destroyed Sri Lankan landscape something out there, easily identifiable and documented. The map was the paradigm that disaster response and reconstruction operations worked within and served (Thongchai 1994, 310; see also Sparke 1998). So I became interested in the tsunami maps as technocratic creations, as productions of nationalism, of a particular imagined community. The maps, produced from on-the-ground and satellite data and coordinates, point toward an effort to organize and assess, to produce and gain knowledge for more efficient and effective national disaster management. While colonial maps represented an island of economic fantasies and aspirations of global power, these maps elicited forms of humanitarian aid and intervention. What was taken for granted in these representations, of course, was the extant nation-form, the geobody of Sri Lanka itself.

Excited to visit the UNOCHA office located in Ampara, the capital of the Eastern Province, I discovered a lone mapmaker sitting in a windowless office, updating numbers in the already delineated borders, lines, divisions, categories, and spaces. On an already existing national template, his job was to update numbers and statistics. Later, as I started my long-term fieldwork four years after the tsunami, the mapping fervor had slowed to a trickle, though as a technology, GIS mapping was integrated into the government's post-tsunami disaster initiatives. The Disaster Management Centre proposed in the government's "Road Map for a Safer Sri Lanka" the creation of a GIS-based disaster risk management system, requiring the coordination and construction of a disaster database to be updated and readily available for public use and mapping.

I do not contest the utility and need for the numerous reproductions of tsunami impact maps; they proved useful in response and reconstruction efforts. At the same time, a long tradition of critical geographers, and feminist geographers in particular, have highlighted the ways in which GIS maps can reproduce and reinforce existing social values and inequalities (Elwood and Leszczynski 2013; Kwan 2002; Schuurman 2000). Moving away from analyses that characterize GIS in dichotomous terms such as "cooptation and resistance" (Elwood 2006) or abstract and real, I want to think further through the reception of and urge to produce these representations of the tsunami. That is, I am interested in the conditions and possibilities of maps, in staying with their trouble (Haraway 2016; Wilson 2017, xi). While maps of conflict zones and regions can be used as political tools or as war by other means, as has been the case in the Occupied Palestinian Territories (Schnell and Leuenberger 2014; I discuss this concept in the next section), the maps of tsunami damage caused little dispute among parties involved in disaster recovery, from the Sri Lankan state to humanitarian agencies. As an act of so-called nature and hence an apolitical force, the damage and destruction wrought by the tsunami did not elicit social or political controversy over territorial depictions of the island. Indeed, as images of mechanical (and reproducible) objectivity (Daston and Galison 2007), these maps of the tsunami's impacts were not *sensational*—at least not in the ways that dramatic tourist-shot videos and news photos were. The maps did not show throngs of people running away from giant waves, or staggering piles of debris. They were, on the contrary, widely promoted for their accuracy, neutrality, and ability to provide information, as immediately as possible, about places in Sri Lanka that were not accessible. The images depicting widespread destruction, death, and displacement could be used to garner the continued support of aid organizations and the sympathy of the global community, and it is true that the tsunami brought in an unprecedented amount of aid to Sri Lanka as well as the possibility of aid-sharing between the LTTE and the government. Hearkening to Strassler's quote at the beginning of this section, each of these maps of the ravaged shorelines of the teardrop-shaped island would, over and over again, also sear into the imagination an objective vision of a united Sri Lankan nation. For indeed, these digitized maps are still "authorised by the nation-state, whose identity is bolstered by them just as the cartographic practices contingently reproduces its instantiation as a sovereign space" (Radcliffe 2009, 440). Nation, like nature, was objectively true.

The immediate aftermath of the tsunami did usher in a renewed desire to rebuild a united nation, where the LTTE and the government of Sri Lanka would come together in the midst of shared devastation. As I mentioned in the introduction, in this spirit, then president Chandrika Bandaranaike Kumaratunga attempted to forge a joint aid-sharing mechanism, the Post-Tsunami Operation Management Structure (P-TOMS), between the government and the LTTE. Though the P-TOMS was eventually signed into law in June 2005, the Sri Lankan Supreme Court issued a stay to block its implementation in July. Just a month later, the Supreme Court announced that President Kumaratunga's term would end in December 2005, following elections in No-

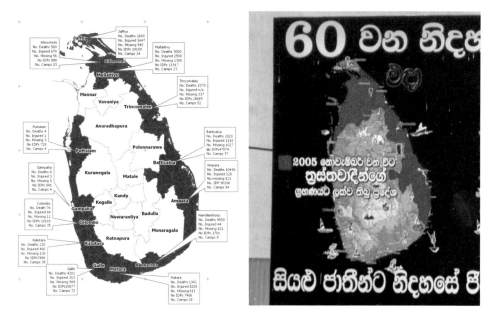

FIGURE 4.6. Visualizing tsunami + war, a juxtaposition.

vember. The effect of the failure of the P-TOMS was to stoke already existing social and political cleavages and fears formed over the decades of civil war. So while the tsunami did result in a brief reprieve in open hostility between the two warring forces, unfortunately, disputes over the failed P-TOMS seemed to cast the final blow to the already tenuous 2002 ceasefire agreement.

Recounting that on the heels of the failed P-TOMS, Mahinda Rajakapaksa's plan, "Mahinda Chinthana: Vision towards a New Sri Lanka," consisted of a new strategy toward the war, I return to the "Year for War" billboard from the introduction, illustrating his vision to "unite" the nation (see figures I.1 and 4.6). In this billboard where "tomorrow" realizes a free Sri Lanka, unification came not through political negotiations and sharing, but rather through military force and the violent defeat of the LTTE.

In the billboard maps and in the tsunami maps—in these metasigns—are the nation. And these maps are emblematic of disaster nationalism, depicting a nation under threat of disaster and in need of disaster—terrorism—management. In the wake of the tsunami, many would refer to the damage as a "war zone," and in some places, where homes and buildings had been bombed or partially destroyed, one development officer claimed that it was difficult to distinguish between tsunami and war destruction. What, then, does disaster *look* like?

A WAR WITHOUT WITNESSES

Four months before the end of the war, in January 2009, as the conflict heated up in northern Sri Lanka, the international human rights community began expressing deep concern for the safety of civilians in the warring zones in the north. The Sri Lankan Army assured the public that great pains were being taken to keep civilians safe, designating "Civilian Safe Zones" or "No Fire Zones." As the army's offensive closed in on Tiger territory, rumors began to flare: humanitarian organizations claimed that civilians were being killed by both warring parties. The government accused the Tigers of using civilians as human shields, while the LTTE in turn reported that the army was shelling indiscriminately, killing innocent civilians and bombing hospitals. Tales and opinions about what was really happening abounded among the international community and throughout the country. In the days leading up to the end of the war, we were interpellated into the government's war efforts by the regular SMS updates on our mobile phones regarding their war progress. For example:

> 5.8.09, 6:32 p.m.
> Previously defined "Safe Zone for Civilians" redemarcated to match current realities: new area, 2km by 1.5km
> –Military Spokesman–

> 5.15.09, 11:12 a.m.
> Rescue mission near completion. 4500 civilians escape LTTE grip, arrive in Govt. controlled areas during the past 48 hours

> 5.17.09, 2:17 p.m.
> Latest update: over 72,000 civilians rescued since Thursday. Remaining LTTE terrorists confined to less than 800 Sq meters.

The Sri Lankan government forbade any international or independent journalists from entering the conflict area, granting access only to the International Committee of the Red Cross, although even they would claim that their aid efforts were stalled due to security concerns several days before the war was declared over. Human rights organizations including Amnesty International, Human Rights Watch, and the UN, frustrated with their exclusion from the war zone and calling the conflict a "war without witnesses," took matters into their own hands. Using satellite data, they published images of the war zone and used them as evidence of artillery shelling and bombing in the Civilian Safe Zone, urging the Sri Lankan government to lay down their arms, come to a peaceful settlement, and open the war zone to humanitarian aid and journalists for the well-being of the civilians.

The area of focus for these satellite images was a small strip of land on the northeastern coast of the island, where the Sri Lankan Army had cornered the LTTE through their military tactic of "attrition warfare" (see figure 4.7). The British Broadcasting Corporation (BBC) was the first to publish these analyses in a news article on May 1, 2009, using images released by the UN. About one week later, the American Association for the Advancement of Science (AAAS) published a report detailing several satellite images, suggesting that they showed evidence of shelling and artillery firing in the no fire zone. The report, made at the behest of Amnesty International and Human Rights Watch, correlated to what news sources were reporting in the area: "shelling . . . in the form of likely shell impact craters and destroyed structures," in evidence "in close proximity to . . . IDP shelters and other structures," and the "obvious removal of thousands of likely IDP structures" from the central part of the CZ ("Conflict Zone") (AAAS 2009, 4, 5) (see figures 4.8 and 4.9).

The Sri Lankan government and the army refuted these allegations. In response, the Sri Lankan Department of Defence produced their own account and interpretations of the satellite images in a document entitled "When the Camera Lies for Terror" (Ministry of Defence 2009). Using their own appointed GIS expert, the government retorted that there was no "ground truth" behind the allegations. According to the government's expert, Ranjith Premalal De Silva, a professor of natural resource engineering and GIS in Sri Lanka: "Most of the analysis was done probably based on object-oriented classification algorithm which interprets an image based on size, shape, colour and other attributes of ground features and the rest of the analysis was purely based on visual interpretations by a group who are totally alien to

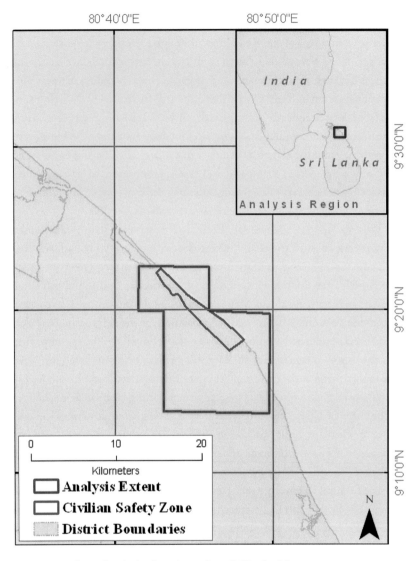

FIGURE 4.7. Area of analysis of war in northern Sri Lanka, May 2009. Source: AAAS 2009.

ground situation in CSZ. Therefore, it is not possible to provide conclusive interpretation based on these analyses without a detailed ground assessment being carried out. Conclusions drawn from the interpretations of these images have no scientific validity" (De Silva 2009). The government's decision to forbid the presence of objective or independent organizations and journalists allowed them to refute claims of damage as mere interpretation and, moreover, scientifically ungrounded. In their analyses of the satellite photos, they argued that the shelling damage could have been done by the LTTE, and that the satellite photos and accompanying analyses were propaganda intended to tarnish Sri Lanka's image internationally. The Sri Lankan Ministry of Defence concluded the report: "This is how the camera is made to lie, giving the lie to the worn out adage that the camera never lies" (Ministry of Defence 2009). Here the Sri Lankan state employed the very same critiques that social theorists have used to question the reality of maps.

I should caution that challenging the veracity of the Sri Lankan government's refutations does not preclude grave concerns regarding the LTTE's war tactics. There is credible evidence too that the LTTE wanted to create a humanitarian disaster, using human shields to serve their own military ends. The exchanges and incidents surrounding these images compel examinations of the trajectory of truth effects and of how political claims are made with imaging technologies. If the Sri Lankan government can employ the same critiques that social theorists use to undermine the truth claims made by images, what is possible to know? With the publication of these images and their respective interpretations, do we actually know more about what was happening in Sri Lanka near the so-called end of its decades-long civil war? In this context, critiques on the basis of objectivity that attempt to undo the god trick do not suffice. The blood through which these images were crafted also stains their intended meanings (Haraway 1988). As such, the content of the images themselves become less significant than the *possibilities* of political truths that are engendered by them. Technological artifacts are always socially and politically situated. It is in these possible competing truths, and in their ambiguities, that specific notions of peace and the limits and capabilities of state power are articulated. Increasing global use of satellite images and maps for humanitarian purposes, as I highlight, does not merely serve state power and control. Critiques of the avowed "perfect" view offered by the map are put to the test with new practices and uses of satellite technology. This is

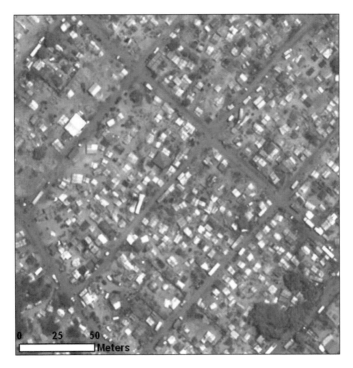

FIGURE 4.8. Satellite image before alleged bombing by Sri Lankan military forces, May 6, 2009. Source: AAAS.

not to valorize satellite and imaging technologies as universally good—but it is to point out that these technologies are not exclusively put to use by the insidious power of the state seeking to coerce and manage territories and populations. Instead, the gaze of surveillance technology compels researchers to understand how possibilities for truth and articulations of social relations are created through visual technologies and representations. Here we can see in action the anthropological insight of Guy Debord, that "the spectacle is not the collection of images; rather it is a social relationship between people that is mediated by the image" (Debord 1994, para. 4).

And so, in the case of Sri Lanka, because journalists and human rights organizations were banned from the safe zone, they found a way in, as it were, by moving up and out. Here, distance was hoped to have the opposite of its usual effect. Rather than purveying a detached, sanitized picture of

FIGURE 4.9. Satellite image after alleged bombing by Sri Lankan military forces, May 10, 2009. Source: AAAS.

the war, distance allowed for a space of radical doubt and for possibilities of ground truth, in an attempt to make concerned citizens of the world witness to human suffering and excessive violence. These efforts reflect the pushes and pulls of sociotechnical practices and relations in satellite imaging. Even if the images are still and frozen moments in time, they are generative of social and material effects. The BBC report mentioned above attempted to galvanize international pressure toward the Sri Lankan government to address the humanitarian plight of an estimated fifty thousand civilians still caught in the war zone. The satellite images, news articles, and human rights reports were also accompanied by visits from the UN humanitarian chief John Holmes and from British and French foreign ministers David Miliband and Bernard Kouchner, all urging the government to at least open a humanitarian corridor, but encouraging a less militaristic resolution to the war. UN Secretary-General Ban Ki-moon called for an immediate halt to the fighting.

FIGURE 4.10. Satellite image detailing bombing in the Civilian Safe Zone, May 2009. Source: AAAS.

Despite these international pressures, Sri Lankan government forces refused to change course. "When the Camera Lies for Terror" was one response. A full-on propaganda machine then countered the international community's narrative, with—among other communications—posters plastered all over the capital city as shown in figures 4.11 and 4.12.[8] Averring its righteousness, while keeping the war zone cordoned off to outside observers, allowed the government to maintain and use situational ambiguity in favor of their war efforts. However, even as the government bolstered its righteousness, its actions evoked more and more global suspicion. If blurring the boundaries between militarization and humanitarianism to embark on the "largest hostage mission in history" was not causing civilian deaths, why wouldn't they grant independent journalists access? Did the Sri Lankan state actually have something to hide?

The "real" content of the satellite images in these exchanges regarding the end of Sri Lanka's civil war are not as significant as the varying interpretations and truths that are spurred by them. Even as the Sri Lankan state asserted its sovereignty as the war ended, its righteousness was also undermined. Speculation and suspicion dogged Sri Lanka's postwar existence. In the aftermath of the war, the UN secretary-general's panel of experts published a report

FIGURE 4.11. Poster supporting Sri Lankan state war efforts. Photo by author.

FIGURE 4.12. Poster supporting Sri Lankan state war efforts. Source: foreignpolicy.com.

FIGURE 4.13. Ban Ki-moon effigy at UN office in Colombo. Source: gettyimages.co.uk.

that found "credible evidence" that war crimes were committed by both the Sri Lankan government and the LTTE (Darusman, Ratner, and Sooka 2011). Given such evidence, they suggested that a war crimes investigation should be launched in Sri Lanka. Before the report's publication, protests erupted outside the now-defunct UN offices in the capital of Colombo, at which effigies of Ban Ki-moon were burned (figure 4.13).

To legitimize their war efforts, the Sri Lankan government conducted their own investigation, the Lessons Learnt and Reconciliation Commission, also released in 2011. The commission was held in the north, where the LTTE was defeated. The thrust of it was to determine why a truce between the government forces and the LTTE broke down in 2002 and to examine the events leading up to the end of the war in May 2009. While the government's report did admit that some civilians were accidentally killed due to the "complexity" of the situation, much to the dismay of international human rights groups, its actions were exonerated through the testimony of its own government and military spokespersons and officials. Two years after the war's end, the British public broadcaster Channel 4 News created a documentary about what "really" happened, evocatively entitled *Sri Lanka's Killing Fields* (see figure 4.14). The

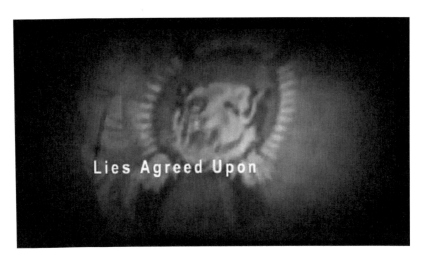

FIGURE 4.15. Still from Sri Lanka's Ministry of Defence film *Lies Agreed Upon*.

informative and revealing about something that we cannot know, or even as something which can lie in the name of political ideology. The various ways in which maps and images are received and interpreted, and used for political purposes, make us privy to the ways in which power works, how "control" of an image's meanings may elicit yet further responses and struggles for power. As Hans-Jörg Rheinberger writes, "It is the nature of these means—material, graphic entities—that they contain the possibility of an *excess*. They contain more and other possibilities than those to which they are actually held to be bound" (Rheinberger 1994, 71). Images do not lead to finality. Disasters do not easily end. As I will further illustrate in the next section, in the case of Sri Lanka, peace became the continuation of war by other means—the means of warring images. Even the state cannot maintain the excesses of disaster nationalism. While seemingly successful at eliminating terrorism in Sri Lanka, in the name of national security, the state still could not put an "end" to the war.

"DEAD" GEO-BODY-POLITICS

For among the most important properties of bodies, especially dead ones, is their ambiguity, multivocality, or polysemy. Remains are con-

FIGURE 4.14. Still from Channel 4 documentary *Sri Lanka's Killing Fields*.

Sri Lankan government's response was swift. They aired their own documentary called *Lies Agreed Upon* (see figure 4.15). This film claimed that Channel 4's documentary was a "total fabrication" created to "tarnish the image" of the island nation. The government refers to the skepticism brought forward by human rights groups as a litany of hatred with ill-concealed contempt against Sri Lanka. The rest of the world, they contended, was merely embarrassed, if not jealous, because they had been unable to defeat terrorism like the Sri Lankan government.

Here is an attempt by the Sri Lankan state to make an "epistemological clampdown," or at least some form of a decisive measure, if only in appearances. After all, as Alan Klima suggests, actual control is not completely necessary: the appearance of control can itself be a real effect (Klima 2002). In these attempts to control the effects and the meanings of the images, the content of the images themselves become less significant than the politics that are engendered by them. It is in these politics of truth-making that specific notions of peace, sovereignty, and the limits and capabilities of state power are articulated (see Feldman 2015; Nesiah and Keenan 2004). Representations and images are able to tell us about the ways in which politics are engendered, enacted, and experienced—these are the "fantastic" elements of visual technologies. Visual technology can be cast into different uses, as something crete, yet protean; they do not have a single meaning but are open to many different readings.

—Katherine Verdery, *The Political Lives of Dead Bodies*

WARNING: GRAPHIC IMAGES IN THIS SECTION.

On May 18, 2009, I found myself at the *kachcheri* (government office) in Ampara, the capital of the Eastern Province in Sri Lanka, for an Early Warning Capacity Building meeting. The meeting was hosted by the United Nations Development Program (UNDP), the International Center for Emergency Techniques, and the National Disaster Management Centre of Sri Lanka. They had invited all local disaster management coordinators to learn more about the new early warning towers that had been erected in tsunami-affected coastal belts around the island. After the meeting, I joined my disaster management friends and colleagues for lunch. As we nibbled on our chicken curry rice packets, all of us received the same text message from the government on our phones: "LTTE leader V. Prabhakaran was killed by troops while trying to flee." Immediately my friend Farood turned the television on, looking for news broadcasts. Deepan went to his computer to check for the latest information available on the internet. This was huge. The ruthless leader of the LTTE, who had led the fight against government forces of Sri Lanka over twenty-five years, was, allegedly, dead. Ten minutes later, we heard firecrackers going off in town. Victory celebrations had already begun. In a flurry of movement, Farood, who worked for the UNDP, urged Deepan to quickly make his way home. Deepan was a young Tamil man; Farood worried for his safety in the predominantly Sinhalese town of Ampara.

I too wanted to go back to my home in Kalmunai, a forty-five-minute tuk-tuk ride away toward the coast. As Farood accompanied me to the central part of Ampara town to help me find transportation home, we encountered boisterous crowds. I watched as trucks full of armed police emptied onto the streets. While my tuk-tuk stopped for petrol, a motorcade led by a truck flying Sri Lankan flags made its way slowly to the main part of town.

Once we got on the road, the fervor faded into the quiet expanse of green paddy fields—Sri Lanka's agricultural belt—and by the time we arrived at Karaitivu Junction to turn onto the A-4, the highway running along the coast, things appeared calm. There was only the usual hustle and bustle of daily life, and no celebrations to be seen. I arrived at my homestay. My homestay family, who obviously had also heard the news, and I huddled around the television, watching the only clear channel they got on their TV: Rupavahini, the state-run channel. A gruesome image of dead Prabhakaran was being looped over and over again (figure 4.16).

His bloated corpse had been pulled from the Nanthikadal Lagoon and laid down on its muddy shores. His army fatigue top was pulled up, exposing a swollen belly. The whites of his wide-open eyes made it seem as if he was startled, as if he had not expected to encounter his own mortality this way. A towel covered the top of his head, which I believe had been blown off, or was in pieces—not completely there, in any case, and probably, I thought while watching, not appropriate to televise nationally. The hands of Sri Lankan Army soldiers swatted at flies swarming around his stiff body. The lagoon water washed a dirty brown, the monsoon sky cloudy and gray.

Disembodied hands of the state once again appeared. They pulled out a wallet, showing an ID with Prabhakaran's photo (figure 4.17). A dog tag with his name. All proof, as if to say LOOK: it really is him. And SEE: he really is dead. In that isolated looping image, there was only him. Would it be a crippling blow to their movement? Awash in dirty brown water, would the LTTE plunge into those murky depths, and could a new united Sri Lankan nation emerge?

Over the decades of Prabhakaran's leadership of the LTTE, he had garnered a kind of "cult of personality" status. Prabhakaran was known to be a ruthless man, famously disciplined. Yet he was also iconic. As Sharika Thiranagama (2020b) writes, "Prabhakaran stands for Eelam as a slogan, Eelam as a map, the LTTE, and himself—as all of those things, and more than them—at the same time." For the Sri Lankan state, killing Prabhakaran was tantamount to eliminating the notion of Eelam, the overplayed looping video an allegory claiming to the public that the Sri Lankan forces had finally rid the island of terrorism.[9] But would that intended political message stick?

The following day, May 19, 2009, the Sri Lankan Army would declare victory over the Tigers and claim an end to the civil war.

The day after Prabhakaran's dead body was made a national spectacle, I made my regular round of visits. An aura of skepticism and solemnity seemed to match the day's grayness. Despite the celebrations happening in other parts of the island and on television, in the east it did not seem an occasion for celebration, and the usually busy streets hosted a somber desolation. Perhaps

FIGURE 4.16. Still from news reports of death of LTTE leader Prabhakaran. Source: *India Today* 2009.

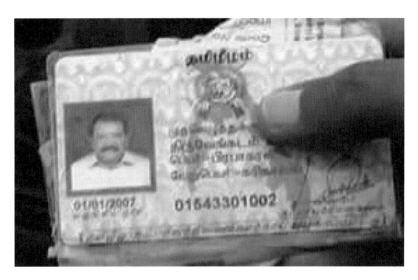

FIGURE 4.17. Prabhakaran's national identity card: Proof, news still, image. Source: Canadian Broadcasting Corporation 2009.

it was because all the shops were closed. A lone Sri Lankan flag waved on a storefront in main Kalmunai town.

In conversations, it became increasingly evident that people were not really convinced of Prabhakaran's death. Lakshmi pointed out that whatever was on the TV was all just "a picture"; it was not necessarily real. Besides, she and Sitamma pointed out, how could Prabhakaran's body have looked so clean-shaven? And did they show his feet? Or the rest of his body? They told me that inside they felt sad. These collective doubts were relayed to me in hushed tones. People could not be public about their disbelief or sadness, for rumor went around about a boy on a bus, who, unaware of the presence of a police officer, proclaimed his disbelief of the death and was unexpectedly smacked and threatened with arrest by the officer for making such a statement. The sadness and doubt should not be mistaken for Tamil nationalism or a devotion to Prabhakaran, but rather signaled the end of one potential channel of advocacy and defense of Tamil minorities in Sri Lanka.

For the LTTE, the death was a crushing blow to their national movement. The group initially denied Prabhakaran's death, although five days later, they too confirmed his demise. And yet even still, photoshopped images of a still-alive Prabharakan watching the looping death-image littered Tamil blogs and Tamil news sites (figure 4.18).

The photoshopped image is an intriguing intersection of what Karen Strassler (2010) has called, in the context of Indonesian photographic practice, "documentary" and "revelatory" history. The continuing circulation of "Prabharakan Still Alive!" serves, following from Strassler, "simultaneously as documentary traces and revelatory signs of presence." It speaks not only to the semiotics of proof but also to the aura of a charismatic individual, and circulation of his image promises "return and recovery" for the LTTE or the possibility of Tamil Eelam (Strassler 2010, 254). As Sharika Thiranagama (2020b) points out, though, this doctored image had a "limited afterlife." Instead, she highlights how more characteristic images of Prabhakaran—images of his face and gaze—continue to circulate in a visual economy of martyrdom and Tamil nationalism.[10]

Susan Sontag writes that "collective memory" is not a remembering but a stipulating: "that *this* is important and this is the story about how it happened, with the pictures that lock the stories in our minds" (Sontag 2003, 86). Surely what was locked in my mind were those death images, that video looping over and over again, Prabhakaran's glazed eyes staring into nothing-

FIGURE 4.18. "Prabhakaran Still Alive!" Source: *HuffPost* 2009.

ness. And how else could it be seen, this death, apart from a triumph? Prabhakaran was a ruthless man; he was a killer, a conscripter of child soldiers, a mastermind, a vengeful leader. Should not everyone, including myself, have been happy? Relieved? The war was finally over. So why were the streets of Kalmunai so quiet? Why did we all stay cooped up inside with the TV on? Why did I feel sad? And why did others tell me that they felt sad and that they didn't believe the pictures? Wasn't the video proof, real evidence of the death? How was seeing actually *not* believing? Though the Sri Lankan state could attempt to hush the voices of dissent, they could not proscribe or control the production of these counterimages. The figure, the geo-body of Prabhakaran that had been presented to the nation and to the world, still incited other forms and modalities of contestation evoking an affective intensity that doubted the demise of the LTTE—or, put differently and more significantly, doubted that the Sri Lankan government had really put an end to the war. The death of Prabhakaran and the end of the war were not seen as necessarily a triumph for everyone in Sri Lanka—rather, Prabhakaran's death incited still more fears that the Sri Lankan state would continue to make life difficult for minorities in the country. My friend Farood came over, and we watched a live telecasting of the Sri Lankan Army in the north, where they had just defeated the Tigers. Celebrating with a show of pomp and circumstance, they fired large cannons into the ocean. Upon viewing this, Farood, a Muslim man with no allegiance to the LTTE, turned to me with tears in his eyes: "Vivi," he said, "you know they were using these to fire upon the people up north."

I discuss these controversies and contestations of these images not to point out that there is no truth, for truths do exist in other forms and narratives, but rather to say that a singular form of truth does not necessarily lead to justice or bring safety, security, and life without fear of retribution in parts of Sri Lanka. In Sri Lanka, the "ground truth," as I have shown, can be wielded by the state as a tool of disaster nationalism, to create a polished veneer of a united Sri Lankan nation, suturing the wounds of the past with a notion of a final resolution. I am suggesting not that our inquiries and interests wallow in a space of many truths, but rather that we must recognize that claims of political truth require much effort to be created and maintained, because they are also continually challenged. By considering images as artifacts made through scientific and political collaborations, agreements, and ideologies, we can see how certain facts in the world (Dumit 1999) are naturalized and come to count as truth. We can look to satellite images and maps as productions of spaces and knowledges that are accepted or not, discredited or not, depending on who interprets and uses them. And perhaps the ways that the world has been partitioned—real and abstract, natural and cultural—are also challenged by these imaging technologies and practices, not because they show us how things *really* are, but rather because they urge us to understand how their articulations gain situational traction and significance (see also Harwell 2000). In Sri Lanka, this is especially the case as the notion of a unified and peaceful Sri Lankan nation and the end of the Tamil Eelam and their respective geo-bodies become reified over and over again. To understand peace as a continuation of war by other means (Foucault 2003; Bhungalia 2023), then, is to recognize that as a social and political state, peace is neither the opposite of war nor the absence of violence, but a state of constant tension, of pushing and pulling, of negotiating and challenging (Klem 2018; Ring 2006). In Sri Lanka, images can uphold ideological pursuits of disaster nationalism in an attempt to preserve the image of the nation, in an attempt to foreclose the possibilities of other narratives, of other stories, of a politics of remembering, but images and their reiterations also can provide a space where contestations emerge (see Buthpitiya 2023), are produced, and have the power to cast radical doubt and open up the possibilities of truth claims.[11]

In the waning days of the war, I sat with Fajiana, a Muslim woman, in her flat, built after the tsunami. She had moved from the north to the east and to Kalmunai in 1990 and converted to Islam; it seemed a safer option for her and her future family. As we chatted, one of her children flipped the television on. The government channel was showing areas of Jaffna, the areas in the north recently liberated by the Sri Lankan Army, the very areas where no one was allowed access. She said she was suffering very much for the civilians up there, including her brother and her brother's children. She had been unable to contact them, and did not know about their situation or whether they were safe. Then something on the TV caught Fajiana's attention. She recognized some of the beachside shots—the area where she used to live, from which she had fled, literally dodging bullets, she told me. She told me her last memories of those beachside homes: they were either shelled and burned out or occupied by Sri Lankan Army forces. But, she said, on the TV of course they were "only showing all very beautiful and nice images."

In May of 2010, about a year after the end of the war, the government proposed to wipe out nearly all LTTE landmarks, bulldozing LTTE cemeteries and the homes of Prabhakaran and other LTTE leaders in the north. The rationale, according to then minister of tourism George Michael, was that the LTTE and war violence should be "forgotten" (de Alwis 2010). War victory memorials lauded the Sri Lankan government's "humanitarian" Sinhala nationalist triumphs, controlling the "official" memories and narratives of the war. Other "war tourism" sites, such as LTTE bunkers and homes of LTTE leaders (which are now closed) and a war museum in the north, also serve to dehumanize the LTTE as a defeated yet still lurking security threat, justifying ongoing modes of militarization and, more tragically, erasing the thousands of civilian casualties and legitimate grievances of Sri Lankan Tamils (Hyndman and Amarasingam 2014).

I found myself returning to this same place, the place where I began this book. It looked like this each time, this shell of a house swathed by the cool lull and breeze of the waves, which, after two years, no longer offered a view of the ocean that led to its partial destruction. An irrevocably altered view. The fisherman who once picked up the sand and told me "This is good land!" still returns. I never did find out who lived here.

I think I can feel the ghosts living among the rock and rubble and the lonely foundation and the protruding metal wires that once supported a house, perhaps, for a family. I can hear the rustle of the palm fronds above, raising the hairs on my skin, a reminder.

Tsunami-damaged home, five years later. Photo by author.

> Many Sri Lankans ask ourselves why, after the end of thirty years of civil war, alongside violence and faultlines that continue to endure, even seemingly *new* violence against *new* actors also takes such predictable forms.
>
> Sharika Thiranagama, "Figurations of Menace"

Postscript

WHEN IS THE POST?

During the summer of 2011, a mysterious and dangerous figure—the "grease devil"—terrorized Sri Lanka. The elusive and fearsome predator attacked at night, assaulting women, keeping scared people inside their homes. The grease devil is, as the name evokes, covered in black grease, making him slippery and difficult to capture. Though his attacks afflicted the entire island, as Rajesh Venugopal observes, "the grease devil provoked widespread acts of public disorder and confrontations with the security forces only among the minorities. It created a rippling crisis of governability in those areas, and particularly in the heavily militarized north and east, many parts of which had only recently returned to government control" (2015, 621). Tamils and Muslims of the north and east were of the opinion that the government was behind the grease devil attacks, and that the attackers were actually military men in disguise. The government refused to acknowledge the grease devil. They cast aspersions on such accusations, proclaiming that "the President does not need any [blood] rituals because he has the people's support.... To suggest that they [the military] would carry out these acts is ridiculous" (Venugopal 2015, 627). About a month after the grease devil started appearing, its appearances gradually petered out. However, the experiences of minority communities in Sri Lanka revealed a pervasive and ongoing anxiety toward state violence—whether the state's absence and inability to manage crisis or its capacity to use excessive force to vanquish it (Thiranagama 2022).

I checked in with my friend to see if things were going all right with her and how things were in Kalmunai. She wrote to me (personal communication, August 31, 2011):

> Yes the situation in Tamil and Muslim areas are very hard. 2 weeks ago very fearful time in my area. But nowaday our area less. And bad in Jaffna and North side of country. Every woman is living with fear. No one open their house after 6 P.M. Men are being [woken] without sleep until midnight. We the Sri Lanka are used to have a fear always. Civil war, Tsunami, White van, now Grease man. Very hard to feel free. 1stly the grease man started their work in Hill country areas, then Eastern, now very badly in Jaffna side. There are too many reasons about Grease man. But all reasons are linked with Government. What to do! This is our luck and this is our part.

I borrow this section heading's question, "When is the post?" from Stuart Hall, who asks: "When was the postcolonial?" (Hall 1996). Hall suggests that "the postcolonial" should compel us to find a productive interrogative space in the notion of the "post." The "post" points toward a shift and movement, in which we can expect aftereffects and emergent power dynamics, discontinuities and continuities, decenterings, encounters. The "post" is productive for rethinking not only the conditions of postcoloniality in Sri Lanka but also those of tsunami and war—the conditions of post-disaster. The "post," then, is not a break from the conditions of the past, but rather ongoing relations with it.

What happens if we consider disasters not just as events, as things that happen to people and to the world, as externalities to be managed, but instead as a structure of power (Wolfe 2006)? In the context of the United States, Christina Sharpe writes, "Transatlantic slavery was and is the disaster. The disaster of Black subjection was and *is* planned; terror *is* disaster and 'terror has a history'" (Sharpe 2016, 5). For her, state-sanctioned killing of Black people has become normative, the "weather," in "this so-called democracy." While Sri Lanka's history and context of course differ from those of the United States, Sharpe's assertion is instructive. In Sri Lanka disaster *is* terror. The disaster of state-sponsored nationalism *is* planned. This is the disaster of disaster nationalism.

In Sri Lanka, the tsunami enabled disaster nationalism: disaster management undergirded by an already existing militarized logic, normalized after decades of war, where tsunami and terrorism would both become institutionalized as threats to national security.

Studying the tsunami and the civil war together *as disasters* illuminates how new technologies and techniques of statecraft—projects of disaster nationalism—can be implemented to further national and political agendas. Indeed, in Sri Lanka, disaster nationalism follows a social and historical trajectory of authoritarian state violence often articulated through Sinhala Buddhist nationalist fantasies.

On the tenth anniversary of the tsunami, in December 2014, I returned to Sri Lanka. It was my first visit since I completed my fieldwork in September 2009. Exiting the airport, I was barraged by eager taxi drivers trying to get my attention, shouting "*ni hao*." After negotiating a taxi, we drove on the recently finished expressway to Colombo. The Chinese Metallurgical Group Corporation was awarded the contract in August 2008 by Mahinda Rajapaksa and commenced construction after the war in October 2009, completing the highway in October 2013.[1] The growing presence of Chinese-built infrastructure and development had changed the greetings that were once "*konichiwa*" to "*ni hao*." Security checkpoints no longer marked major throughways; instead, new constructions greeted me. Festively decorated for Christmas, a new shopping arcade near Independence Square sparkled. I noticed new restaurants and high-rise flats and numerous construction projects. Colombo's roads had changed—one-ways that did not exist before confused my sense of direction. This was postwar Sri Lanka.

Yet, beneath this shiny postwar facade, as Harini Amarasuriya and Jonathan Spencer (2015) highlight, is an anxious Sri Lanka; belying these "beautification" projects and infrastructural and economic developments in Colombo is the continued militarization of the island. In 2015, after the reelection of Mahinda Rajapaksa, the Urban Development Authority was formally placed under the Ministry of Defence, led by the president's brother, Gotabaya. Gotabaya's vision of Colombo reflects his aggressive militarism— or his "short-term expediency" (Amarasuriya and Spencer 2015, S71)—and furthermore, his attachments to Sinhala Buddhist nationalism. Colombo has never been a predominantly Sinhala Buddhist space, but a "place of minority life," dense with multiple ethnicities (Thiranagama 2011). The spaces of Gotabaya's "development" projects have "relocated" Muslims where Sinhalese are the minority (Amarasuriya and Spencer 2015). The end of one virulent nationalism, the LTTE, continues to strengthen and embolden Sinhalese Buddhist nationalism (see also Fernando 2013). Disaster nationalism endures.

During the same visit in 2014, I left the sparkling city to spend the Christmas holiday on the east coast. That December, Sri Lanka was doused with monsoon rain, which led to flooding and landslides across the island. As with the tsunami, the Eastern Province was the worst affected. According to a briefing note from ACAPS, an independent information provider for humanitarian needs, the Sri Lankan Disaster Management Centre (DMC) reported that fewer than 100,000 people were affected as of January 6, 2015, with 5,700 staying in evacuation centers. The briefing note also pointed out a discrepancy between the government's reports and the United Nations Office for the Coordination of Humanitarian Affairs (UNOCHA) report, which gave higher numbers: 1.1 million people affected and 30,000 people hosted in 230 evacuation centers, with 120,000 people displaced, the majority staying in 618 safety centers and others taking shelter with family or friends (ACAPS 2015). The official Sri Lankan Government News Portal (news.lk) also provided updates of the government's coordinated efforts. Sri Lankan Army troops, under the Security Force Headquarters of the East (as with other affected provinces on the island), were tasked with rescue and relief operations (news.lk 2014). After my research assistant and I reunited, we waded through massive puddles and flooded roads to visit old friends on the anniversary of the tsunami.

We first ran into Selvi, mother of "little cow." She was renting the small stall in front of the tsunami flats, where she continued to manage her neighborhood shop. We sat under a metal awning, talking loudly amid the roar of rain pounding metal and as sheets of rain poured around us. The dark skies seemed appropriate as she once again recounted memories of her little cow. She showed us a letter she received from a local orphanage thanking her for her donation of food and clothes. Her older daughter, still saddened by memories of the tsunami, did not want to stay near their old home and instead continued to stay with family up north, in Batticaloa. Around us, there were not any major commemorative events; it seemed that most people conducted private remembrances, visiting local shrines and cemeteries. The heavy rains were also a deterrence.

Early that same morning, I had watched the televised National Safety Day exhibition in Hambantota, Mahinda Rajapaksa's hometown, in the conference hall named after him. The theme for the exhibition was "Safety Sri

Lanka—Let Us Unite to Eradicate Disasters" (see also Benadusi 2020). A moment of silence was supposed to be held, as is done yearly, from 9:25 to 9:27 a.m., the time when the tsunami crashed into Sri Lanka's shorelines. But the exhibition too was delayed; the DMC was also trying to attend to the torrential rains that had caused island-wide flooding. At the exhibition, each government unit under the purview of the Ministry of Defence, including the Department of Meteorology, the Urban Development Authority, the DMC, and the National Building Resource Organization, had an exhibit. The DMC had built a mini replica of a disaster warning tower, made more immersive with recordings of the siren.

On this return trip to the east, I noticed the conspicuous absence of the checkpoints that used to dot the A-4 highway. People relayed to me that the security situation was "more relaxed," though according to a friend and local administrator the number of military camps had actually increased—it was just that they were more integrated into the communities and civil society (see Fernando 2020; on "post"-war life in the east, see also Spencer et al. 2014). I saw that the site of the temporary housing scheme discussed in chapter 3 had been turned into a Sri Lankan Navy camp. I was discouraged from taking a photo by my research assistant as we walked past it. We did not want to draw any suspicion, she reminded me. While I have not been able to confirm that the number of military camps has increased in the east, studies outlining military and defense spending show that since the end of the war, Sri Lanka has spent more on its defense budget, even more than at the peak of the war in 2009. The bulk of these funds go toward the army: 2.5 percent of Sri Lanka's labor force is in the armed forces, a disproportionately high number, considering the end of the war (Wickramasinghe 2020; see also Alphonsus 2021).

What national security concerns, what "figurations of menace" (Thiranagama 2020), persist in this ongoing and ubiquitous statecraft of militarization and securitization, under these ongoing conditions of disaster?

Immediately after the end of the war, Mohamed, a Muslim fisherman struggling to earn a living for his family in the east, told me that if Sri Lanka is "one nation" or "one country" as the president says, that means it is a Buddhist country. Even though the war was with the LTTE and Tamils were targeted, as minorities, he said, Muslims were impacted too; the president with his military might could "destroy" people. He worried about the possibility of Sinhalization of the east; he worried about the future for his children. Mohamed's concerns were prescient. Since the end of the war in 2009, Muslims increasingly have become the next minority to bear the brunt of the violence of Sinhala Buddhist nationalism. In particular, the Bodu Bala Sena (BBS or "Buddhist Power Force"), an ultra-Sinhalese Buddhist nationalist organization helmed by monks, has organized and continues to organize protests and campaigns against Christians and especially Muslims, as potential threats to Buddhism in the country. In a speech, BBS leader and monk Galagoda Aththe Gnanasara declared, "We the clergies should aim to create a Sinhala government. We will create a parliament that will be accountable for the country, a parliament that will protect Sinhalese" (Aneez and Shah 2019).[2] For according to him, the Sinhalese are the "historical owners of the country." Anti-Muslim actions have included the call to abolish halal certification, the desecration of mosques, vigilante and communal violence, and anti-Muslim protests and riots. Most of these incidents have gone unpunished (Holt 2016; Sarjoon, Yusoff, and Hussin 2016; see also Haniffa 2015a).

My 2014 visit to Sri Lanka also bore witness to a political change in the making. Under his self-appointed constitutional right, Mahinda Rajapaksa, who had removed term limits for the country's presidency by passing the Eighteenth Amendment to the Constitution, called for early elections in November 2014. Emboldened by war victory hubris, he mistakenly predicted that, as in the previous election, he would easily win. In a surprising turn of events, he was defeated by the opposition candidate Maithripala Sirisena. Sirisena, who called himself "the common candidate," won with the overwhelming support of minority votes, especially from the north and east, as well as the votes of the many Sinhalese fed up with and alienated by Rajapaksa's authoritarian regime of cronyism and family favoritism. Sirisena pledged to repeal Rajapaksa's Eighteenth Amendment and his parliamentary majority, rescinding the near-authoritarian executive power and reintroducing important constraints and balances of power. His ambitious and promising reform agenda also included reining in corruption and abuse of power, a new constitution, the repeal of the Prevention of Terrorism Act, and agreement with the UN Human Rights Council (UNHRC) to investigate war crimes and crimes against humanity by all parties involved in the civil war.

By these measures, Sirisena's government resoundingly failed. A new constitution never materialized; the PTA is still in place; military occupation of Tamil and Muslim lands and properties in the north and east continued unabated. Sirisena refused to punish soldiers alleged to have committed war crimes and failed to honor transitional justice promises made to the UNHRC, saying, in language echoing his predecessor's, words that he did not want to "re-open old wounds" (Colombo Correspondent 2019).

Then, in October 2018, Sirisena sacked Prime Minister Ranil Wickremesinghe and instated the former president he defeated, Mahinda Rajapaksa. The courts struck down this power move and reinstated Wickremesinghe as prime minister. Instead of gaining Rajapaksa's support for reelection, as he might have hoped, Sirisena found himself competing with the ex-president's brother, former Minister of Defence Gotabaya, as a presidential candidate. The tragic bombing of Colombo churches in April 2019 dashed any hopes for Sirisena. Allegedly carried out by local members of ISIS, the "Easter bombings" flagged serious national intelligence and security failures. The bombings also stoked anti-Muslim sentiments while reenergizing Sinhalese Buddhist nationalist fervor—Sirisena pardoned the aforementioned BBS leader, Gnanasara, who had been jailed in 2016 for contempt of court—and paved the way for Gotabaya Rajapaksa's bids for absolute executive power.

As the defense minister who defeated terrorism, Gotabaya "The Terminator" Rajapaksa convinced voters that he was the only presidential candidate who could maintain Sri Lanka's national security. In November 2019, carried by the votes of his ultranationalist Sinhala Buddhist constituency, Gotabaya Rajapaksa and his platform of "Vistas, Prosperity and Splendour" won the presidential election. He appointed his brother and former wartime president, Mahinda Rajapaksa, prime minister. As Sharika Thiranagama (2020a) describes, "The civil war saw the complete militarization of government. The Sri Lankan Army had always been subordinate to the president. But now, Gotabaya Rajapaksa, as a former army man and as the president, symptomizes a collapse that has been a long time coming."

PANDEMIC NATIONALISM

While today the threat of another tsunami may seem distant, the management of other disasters such as flooding, drought, and landslides—the most commonly occurring disasters in Sri Lanka—continues to draw upon the manpower and direction of the armed forces. And eleven years after the end of the war, in the midst of the COVID-19 pandemic, "those at the forefront in playing key roles in the fight against COVID-19 are the same people who fought a winning war 11 years ago" (Weerakoon 2020). With the pandemic likened to a war and framed as a "national security challenge," COVID presented yet another opportunity to further militarize social life (Fonseka 2020; Fonseka and Ranasinghe 2021). For example, former and present military officials were appointed to positions in government including the posts of Secretary to the Ministry of Health and Director General of the Disaster Management Centre. In particular, the vague yet expansive role of the two government entities created by President Gotabaya Rajapaksa tasked with coordinating and overseeing the country's handling of the pandemic were concerning:

1. The National Operation Centre for Prevention of COVID 19 Outbreak (NOCPC) which was created to "coordinate preventive and management measures to ensure that healthcare and other services are well geared to serve the general public" (President's Media Division 2020) and is headed by Lt. Gen. Shavendra Silva, the Commander of the Sri Lankan Armed Forces.
2. The Presidential Task Force, established to "Direct, coordinate and monitor delivery of continuous services and for the sustenance of overall community life" (President's Media Division 2020) and is headed by the president and prime minister's brother Basil Rajapaksa.

Rather than using existing constitutional and legal structures, such as the Disaster Management Act of 2005, or a directive by the National Council for Disaster Management, the president appointed not medical or public health officials or experts but a former commander of the Sri Lanka Army who is currently facing charges of war crimes related to the end stages of the civil war in 2009 and is banned from travel to the United States.[3] The appointments of Lt. General Silva and Basil Rajapaksa are part of a broader trend of President Gotabaya Rajapaksa's "rule by committee and taskforce," exemplifying the centralization of executive power and vulnerability of democracy in Sri Lanka (Fonseka and Ranasinghe 2021; Fonseka, Ganeshathasan, and Welikala 2021)—the collapse Thiranagama describes above. With this complete militarization of the government, then, the Disaster Management Act, as a mechanism of securitization, was not even necessary to invoke to justify presidential actions of pandemic nationalism.

In fact, at the beginning of the pandemic in March of 2020, in an attempt to build a parliamentary majority with new elections, Gotabaya Rajapaksa dissolved Parliament. During this dissolution, he created two task forces in addition to the COVID-related task forces: the Presidential Task Force to Build a Secure Country, Disciplined, Virtuous and Lawful Society; and the Presidential Task Force for Archaeological Heritage Management in the Eastern Province. Like the COVID task forces, these are composed of military and law and order officials. And, without Parliament, there was no legislative oversight over executive actions (Peiris 2021): "COVID-19 has provided the government the perfect excuse for effective Executive aggrandisement and militarisation" (Saravanamuttu 2021, vii). The Rajapaksa government has sought to strengthen their ethno-religious nationalist dominance and ideology through the centralization of executive power, continuing to sow hardship and fear among Sri Lankan minorities (Fonseka and Dissanayake 2021). On the ground, the militarized COVID response—army-run checkpoints, roundups, quarantine centers, and contact tracing efforts—stirred fear, especially in the minority north and east of the island. Building off the Easter attacks, the government intimidated and targeted Muslims with disproportionate scrutiny of COVID infections in Muslim neighborhoods, in addition to social media disinformation campaigns blaming Muslims for spreading the virus (see Moinudeen 2021). Most egregious was the enforcement of mandatory cremations, rather than allowing Muslims (and Christians) to bury their dead in accordance with religious customs. Though the World Health Organization (WHO) guidelines permitted burials and the Sri Lankan Medical Association reiterated that burials posed no threat of viral spread, the Sri Lankan government, as they did with satellite images during the war, used their own medical and academic experts to allege otherwise. The ban on burials was only lifted in February 2021, but in keeping with the ultra-Buddhist nationalist agenda, the government designated a remote island, Iranathivu, populated by Tamil returnees, as the first site where burials could take place.

Disaster once again was exploited by the powerful structures of the militarized nationalist state. And Sri Lanka was not alone in exploiting the pandemic to serve nationalist ends. Somewhat early in the pandemic, in March 2020, the United Nations urged states to avoid overreach of security measures in response to the COVID outbreak: emergency measures "should not function as a cover for repressive action under the guise of protecting health" (UNOHCHR 2020). Shortly after, in April 2020, UN Secretary-General António Guterres cautioned that the COVID-19 pandemic was "fast becoming a human rights crisis," with the risk that it could "provide a pretext to adopt repressive measures for purposes unrelated to the pandemic"

(UNOHCHR 2020). In a special issue of *Nations and Nationalisms* on COVID-19, several authors highlighted how the pandemic had amplified nationalisms and exacerbated ethnic and nationalist conflicts (Woods et al. 2020). The United States enacting wartime decrees such as the Defense Production Act, Hungary granting its prime minister indefinite emergency power, and the Israeli Security Agency's electronic tracking system—previously used for anti-terrorism purposes—to monitor citizens' movements are just a few examples in which technologies of pandemic nationalism can be seen. In India, conspiracy theories blaming Muslims abounded, as they did in Sri Lanka, dangerously pushed by Hindu nationalists.

Thinking more broadly, and given the anticipated disasters attributable to climate change, there is no place in the world today that is not facing future disasters. Indeed, the disasters are already here. It is crucial, then, to carefully examine how states can manage and respond to them responsibly.

These pages have been critical of a particular type of disaster management and particular ways of defining and understanding disaster, preparedness, and safety. This critical approach is not to suggest that disaster preparedness and response efforts and programs should not exist. It asks instead whether it is possible to divorce safety and disaster preparedness from (national) security. For, as those living in eastern Sri Lanka can attest, state securitization and militarization of disaster management, preparedness, and reconstruction may offer safety only for select communities and create new or exacerbate already existing forms of precarity and vulnerability.

Each of the preceding chapters has moved through the multiple, syncopated, and sometimes undisciplined tenses of, modalities of, and orientations to disaster in Sri Lanka, challenging the cyclical linearity of disaster and risk management teleologies. I have shown how the future-oriented, preemptive structuring logics of disaster risk management coevolve with strategies and logics of national security—disaster nationalism—and lead to ongoing insecurity and precarity. As my friends and interlocutors showed me, threat is not just the possibility of another tsunami or even outbreak of war but also the enduring power of the state in the guise of national security. They also showed me glimpses into their lives, their anticipations, their endurances that disrupt this fantasy—their interrogations of the "post."

Notes

PREFACE

1. For more detailed information on these events, see, for example, Centre for Policy Alternatives (2023); Fedricks et al. (2023).
2. For example, a major blunder was summarily banning chemical fertilizers, causing a major drop in agricultural production, especially of rice. Gotabaya Rajapaksa also implemented a major tax cut after his election, further shrinking the country's economic reserves.
3. As Ahilan Kadirgamar (2022) writes, Sri Lanka's economy has been crisis-prone since liberalization.
4. This and many other dimensions of the Aragalaya were discussed by speakers Swasthika Arulingam, Marisa De Silva, and Farzana Haniffa in the virtual panel "The People Revolt: Sri Lanka Panel," hosted by the University of California Santa Cruz Center for South Asian Studies, October 11, 2022. Available on YouTube at https://youtu.be/9VAK2PaOaZM?feature=shared.

INTRODUCTION

1. On militarization in Sri Lanka, see de Mel (2007). As Catherine Lutz (2002) proposes, militarization is more than an intensification of resources toward military purposes; it is also the shaping of institutions congruous with military goals.
2. See Remes and Horowitz (2021) on disaster as analytical conceit.
3. With thanks to Kim Fortun for this challenge.
4. In expanding the boundaries of what might be included by "disaster" and, in particular, understanding forms and techniques of governance as disaster, I also am thinking with Adi Ophir's (2010) theorization of

discursive catastrophization, in which he uses "man-made" disasters to revise concepts of political theory and the "catastrophization" of the Occupied Palestinian Territories and contemporary zones of emergency. See also Vázquez-Arroyo (2013).

5. There is a growing body of critical scholarship on disasters, especially over the last two decades; for example, Fortun (2001); Petryna (2002); Dowty and Allen (2010); Adams (2013); Bond (2013); Knowles (2013); Kimura (2016); and Morimoto (2023).

6. The idea of disaster as method draws on the work of Knowles and Loeb (2021), who employ disaster as method in their interscalar work examining the "voyage of the *Paragon*" within the broader (slow) disaster of Hurricane Harvey.

7. The Department of Census and Statistics completed the 2012 census for the entire country—for the first time since 1983. Because of the war, the 2001 census only surveyed eighteen of twenty-five districts in the country. Unsurprisingly, the missing districts were in the north and east (Department of Census and Statistics 2012).

8. In providing this abbreviated history, I do not intend to suggest a singular or monolithic Sinhalese identity. Rather, I show how ethnicity is instrumentalized in national politics.

9. Following independence in 1948, and the passage of several Parliamentary Acts, Tamil representation and political power were reduced. In particular, one of the Acts made Indian Tamils "non-citizens" in Ceylon, thus reducing the Tamil minority. According to Hoole et al. (1990), elite Tamils were willing to disenfranchise their lower-class Tamil counterparts in order to preserve their status in politics, a deal made through personal guarantees by the prime minister of Ceylon at the time, Don Stephen Senanayake. The leftist parties, such as the Trotskyists, the Bolshevik-Leninists, and the Communist parties, were all vehemently against the disenfranchisement of the Indian Tamils. Though Kumari Jayawardena (2003) points out that the progression of Sri Lanka's early democratic politics could be better understood through class interests rather than through ethnic ones, a Sinhala majority would come into power.

10. On the development of Tamil political consciousness, nationalism, and the LTTE, see *The Broken Palmyra* (Hoole et al. 1990), *Sri Lankan Tamil Nationalism* (Wilson 2000), and *Learning Politics from Sivaram* (Whitaker 2007), which also highlight different militant Tamil nationalist groups beyond the LTTE.

11. The 1987 Marxist JVP insurrection was the second anti-state revolt on the part of this movement, which initially began in the early 1970s to address the economic struggles of rural youth. After being ruthlessly put down by the Sri Lankan state in the late 1980s, the movement later

reemerged as a viable electoral party in the 1990s. The late '90s and early 2000s saw the JVP embrace and ally with Sinhala nationalist ideologies and groups, drawing support by mobilizing class and economic anxieties and precarities through nationalism (Venugopal 2010; see also Moore 1993; Uyangoda 2008).

12. The Indian Peacekeeping Force (IPKF) was formed under the 1987 Indo–Sri Lankan Accord. In 1987, president of Sri Lanka J. R. Jayawardena and Indian prime minister Rajiv Gandhi signed the Indo–Sri Lankan Accord, which was an effort to end the conflict. Under these conditions the Thirteenth Amendment to the Constitution established the devolution of powers to Provincial Councils, and the IPKF would be sent to Sri Lanka to enforce the ceasefire and to disarm the LTTE. The Accord was not well received, especially by Sinhala nationalists including the JVP, who waged an anti-state insurgency against the government soon violently put down by the state, led at the time by Jayawardena's predecessor, Ranasinghe Premadasa. Once the JVP had been quelled, Premadasa secretly colluded with the LTTE to push the IPKF to leave Sri Lanka. In the north and east the LTTE fought the IPKF, harming civilians in the crossfire. The IPKF's response led to hundreds of deaths, arrests, and rapes, and eventually to the group's withdrawal in 1990. This retreat allowed the LTTE to consolidate its power in the north and east (Hoole et al. 1990; Shastri 2009).

13. Note that, while this was a Parliamentary meeting on "natural disasters," the document covers more than just "natural" disasters—as does the ensuing National Disaster Management Act. The terms *disaster risk management* (DRM) and *disaster risk reduction* (DRR) can be used interchangeably, but it is also possible to consider DRM as the application or implementation of principles of DRR.

14. See Choi (2009) for a discussion of risk management rationale and technology after the tsunami in Sri Lanka.

15. Risk reduction had in fact been promoted by the United Nations in the 1990s, according to the United Nation's International Strategy for Disaster Reduction (UNISDR). In this approach, "there are no such things as natural disasters, only natural hazards" (UNISDR 2019). Disasters happen after a natural hazard strikes. The impetus for disaster risk reduction is to reduce the damage caused by natural hazards like earthquakes, floods, droughts, and cyclones through an ethic of prevention. National governments are primarily charged with the task of disaster risk management (DRM, another term for DRR). This includes the broad development, application, and assessment of policies, strategies, and practices to minimize vulnerabilities and disaster risks throughout society. Whereas previous systematic approaches to natural disasters focused primarily on responses to them, beginning in the 1990s (designated by the UN as an

International Decade for Natural Disaster Reduction), early conceptions of disaster risk management were initiated in a "global culture of prevention." In 2000, following the International Decade for Natural Disaster Reduction, the United Nations created the International Strategy for Disaster Reduction as an official interagency task force and interagency secretariat (UNDRR, n.d.).

16. For a discussion comparing how post-tsunami international aid and the presence and practices of international nongovernmental organizations (INGOs) factored into, on the one hand, "peace" between the Free Aceh Movement/GAM (Gerakan Aceh Merdeka) and the Indonesian government and, on the other, an increase in hostilities between the LTTE and the Sri Lankan government, see de Alwis and Hedman (2009); Enia (2008).

17. On the tsunami and failure of the P-TOMS as a "missed political opportunity" for peace, see Fernando (2015).

18. For a more detailed analysis of the P-TOMS, see Keenan (2010).

19. For more on post-tsunami recovery efforts in Sri Lanka, see McGilvray and Gamburd (2010).

20. Much of the literature on conflict and disaster focuses on whether natural disasters lead to peace or to more conflict, examining the causal relationships between conflict and disaster. Yet, as Peters and Kelman (2020) caution, causal frameworks obscure the complexities of both disaster and conflict; disasters do not create "new" conflicts or peace, but rather reproduce or rearrange political processes of conflict and peace.

21. Elsewhere, I have discussed the complexities of my race and "whiteness" while conducting fieldwork as a broader, persistent issue within anthropology (Choi 2020). On conducting fieldwork as a woman of color, see also Funahashi (2016); Navarro, Williams, and Ahmad (2013). On the dangers and violence toward women of color, see Berry et al. (2017); Williams (2017); and, in a Sri Lankan context, Jegathesan (2020).

22. Thiranagama describes postwar violence and militarization in Sri Lanka as ambiguous, dispersed, and ever-present, such that "the state and its militarised life search for and mobilise around an *object* of menace—Tamil and Muslim minorities, the LTTE, COVID-19—while its *solution* to national menace remains the same: more militarisation" (2022, 202). On the state and fantasy, see also Aretxaga (2001). On fantasy and disasters, see also Clarke (2001); Liboiron and Wachsmuth (2013).

23. The *New York Times* (2010) effuses, "The island, with a population of just 20 million, feels like one big tropical zoo: elephants roam freely, water buffaloes idle in paddy fields and monkeys swing from trees. And then there's the pristine coastline. The miles of sugary white sand

flanked by bamboo groves that were off-limits to most visitors until recently are a happy, if unintended byproduct of the war."

24. See also Jackie Orr (2006) on how governance of the future can become "psychopolitical."

25. As a related aside: Greenhouse also argues that anthropologists remain stuck in the time of the nation-state, what Wimmer and Schiller (2002) call out as "methodological nationalism." I agree with this diagnosis, though I do not tackle it in this book. However, not unrelatedly, my effort is to trace and critique the formations of nationalism and to illustrate the effect of such persistent forms of violence, rather than to assume the nation as a primordial or ontological social thing.

26. On nationalism in Sri Lanka (and gender), see de Alwis (1995); Jeganathan and Ismail (1995); Rogers (1994); Spencer (1992, 2003); Whitaker (2007). On nationalism in other contexts, see Aretxaga (1999); Chatterjee (2004, 2005); Puar (2007); Verdery (1993).

27. On "thinking against crisis logics," see Elinoff and Vaughan (2021).

28. The temporality of the 2004 tsunami, as Helmreich (2006) finds, includes tensions and intersections of different parsings of time and history as they pertain to nature, bureaucracy, science, and emergency.

29. For some examples, see Daniel (1996); Das et al. (2001); Jayawardena and de Alwis (1996); Jeganathan (2000); Nesiah and Keenan (2004); Ruwanpura (2006, 2009); Scheper-Hughes and Bourgois (2004); Tambiah (1992, 1997); Visweswaran (2013).

30. On violence manifesting in more ordinary ways, see Ralph (2014) and Stewart (1996, 2007); see also Ahmann (2018); Coronil and Skurski (2005); Nixon (2013). On crisis, see Berlant (2011); Dillon and Lobo-Guerrero (2009); Moten (2003); Povinelli (2011); Roitman (2013); Vigh (2008).

31. With thanks to Seulgi Lee for this insight.

32. This is inspired by Shannon Dawdy's (2006) "taphonomic" approach to sifting through post-Katrina rubble and detritus. Borrowing from Foucault's *Archaeology of Knowledge*, taphonomy is the archeological record, which is not just a reflection of social process but is a social process itself.

1. EMERGENCE

1. Notably, in a speech during a visit from US president Ronald Reagan in 1984, then president J. R. Jayawardena proclaimed: "Sri Lankan nation has stood out as the most wonderful nation in the world because of several unique characteristics. Sinhala nation has followed one faith, that is Buddhism for an unbroken period of 2500 years . . . there is no

other nation that can boast such a heritage.... Another unique heritage is the country's history of sovereignty and territorial integrity. No other nation has enjoyed national independence for such a length of time as we have.... We are the most wonderful nation in the world. We must be proud of our history" (quoted in Krishna 1996, 309).

2. With the help of friends and my research assistant, I was able to locate some published accounts documenting the effects of the cyclone in Tamil (though even a search at the Batticaloa library was not fruitful!), and Dennis McGilvray provided me with his extra copy of Vini Vitharana's (1990) short booklet *The Night of Doom*, an account of a government administrator reliving the night of the cyclone in Batticaloa. I found only one account in *The Broken Palmyra* (Hoole et al. 1990), which mentions the discrimination that Tamils in the east and in Parliament felt in the provision of cyclone relief, adding to what the book details as "The Development of a Tamil Political Consciousness"—an astute observation, given the developments happening in Sri Lanka politically.

3. In 1990, the LTTE expelled Muslims from the north. In Jaffna, Muslims were given two hours to leave, forcing them to abandon their belongings. Muslims in other districts in the north were given a forty-eight-hour ultimatum (Hasbullah 2016). Most settled in Puttalam, where they struggled to rebuild their lives (see Haniffa 2008; Hasbullah 2001; Thiranagama 2011). The end of the war has been fraught with challenges in regards to Muslim reconciliation and resettlement in the north (see Haniffa 2015a; Yusoff, Sarjoon, and Zain 2018).

4. On the impacts of language policies in education in Sri Lanka, see Davis (2020).

5. In a moment when transnationalism, globalization, and finance capital questioned the significance of nation-states and, hence, the notion of "nation," Katherine Verdery instead suggested: "The most comprehensive possible agenda for the study of nationalism is, therefore, the study of historical processes that have produced a particular political form—nation-states—differently in different contexts, and of the internal homogenizations that these nation-states have sought to realize in their different contexts" (1993, 43).

6. Though, as Patrick Peebles (2001) points out, the British did not count themselves in that category.

7. For more on the disenfranchisement and exclusion of Malaiyaha Tamils in postcolonial minority politics in Sri Lanka, see Jegathesan (2013, 2019).

8. For accounts of abuses under the PTA, see Human Rights Watch (2018a).

9. See Human Rights Watch (2018b) and, more recently, the Centre for Policy Alternatives' (CPA) Petition against the Anti-terrorism Bill of 2023 (which has been tabled; Centre for Policy Alternatives 2024).

10. Even animals can be abstracted into homogeneous symbols in service of exclusionary nationalist policies. As Radhika Govindrajan (2018) shows, "cow protection" as a "cornerstone" for Hindu nationalism—and a justification for violence toward Christians and Muslims—in India encountered obstacles because it produced an abstract and homogeneous symbol of the cow, subsequently erasing the complex relational beings they are to the people who live with them.
11. In a similar vein, Kalinga Tudor Silva outlines how malaria eradication and control of the Dry Zone—an area with a large Sinhala peasant population—also became a tool of the political elite, as part of a postcolonial development and economic agenda, also driven by a nationalist agenda (Silva 2014).
12. Relatedly, Banu Subramaniam (2001) has argued that racialized and xenophobic rhetoric describing "exotic" plants as "invasive species" and "aliens" is symptomatic of nationalist anxieties of globalization, immigration, and changing demographics and gender norms in the United States.
13. On the vivisectional violence of the modern state, see Visvanathan (1997).

2. ANTICIPATION

1. This was the goal as outlined in 2009, when I was in Sri Lanka. In total, seventy-seven warning towers were built. As of December 2023, fifty-seven of these towers were not functioning (Nizamdeen 2023).
2. Andrew Lakoff and Stephen J. Collier (2015) build on Michel Foucault's (2007) work on disciplinary biopower to discuss how "vital systems" rather than population become the focus in preemptive techniques of disaster management. They suggest that apparatuses of vital systems security are a form of reflexive biopolitics that have emerged in response to the risks of modernization (see also Beck 1992). In light of new modes of networked technologies of power, I would add to this discourse Gilles Deleuze's (1992) suggestion that we are moving beyond a disciplinary society into a society of control.
3. See Winslow and Woost (2004) on how unending war became the grounds upon which to fashion lives. As the war grew into its own reality, its reproduction could be less and less tied up with the politics of ethnicity and more dependent on the politics of profits and making a living.
4. Celia Lowe (2010) offers a different perspective on how biosecurity plays out in Indonesia, where "vital systems" and infrastructure differ markedly from their counterparts in the United States.

5. On the anticipation of danger and insecurity on checkpoints in Sri Lanka, see Jeganathan (2000) and Thiranagama (2011); see also Pieris 2015. Scholars of technology and technocracies teach that the construction and implementation of infrastructures and technologies reveal much about politics, state power, and nationalism (Alatout 2008; Mitchell 2002; Pritchard 2010; Scott 1998). On these dynamics in Sri Lanka, see Tennekoon (1988).
6. That the LTTE nationalists proclaimed the northern and eastern parts of the island as solely for Tamils problematically suffused a monoethnic imagination with the territory, thus eliding the long history of both Tamil and Muslim settlement and sociality in those regions (McGilvray 2008). The coastal part of the eastern district is majority Muslim and Tamil, while the interior of the district is predominantly Sinhala-Buddhist, and much has been written on the national redistricting policies and government resettlement projects in the area (Hasbullah and Korf 2013; McGilvray and Raheem 2007; Spencer 2003). Chapter 1 discusses this in more detail.
7. When I left in late September 2009, Sri Lanka's post-conflict situation remained tense. The State of Emergency had just been extended by Parliament; Tamil civilians were still interned in camps in the north, and the government had increased the military budget by 20 percent and increased the size of the military by 50 percent, all in the name of protecting Sri Lanka's fragile "security." Since the war, Sri Lanka's defense budget has increased each year. By 2023, defense accounted for nearly 10 percent of public spending, while security force personnel made up half the government's salary bill (Aljazeera 2023).
8. See Reddy (2020) on how a warning for an earthquake in Mexico that never arrived can lead to distrust in alert systems.
9. The state of emergency was finally lifted in Sri Lanka on August 31, 2011, but the government has retained some of its emergency powers by consolidating them into existing Public Security Ordinances and the Prevention of Terrorism Act. For a legal explanation, see Weliamuna (2011).
10. Elsewhere I have written about how the complex layers of disaster have created certain "infrastructures of feeling" that have become constitutive of everyday life in the east (Choi 2021).
11. Notable examples include the assassinations of Lasantha Wickrematunga and Sivaram "Taraki" Dharmaratnam, the disappearance of Prageeth Eknaligoda, and the arrest of J. S. Tissainayagam. For a poignant and arresting portrayal of journalism, nationalism, and the Tamil movement, see Mark Whitaker's (2007) biography of Tamil journalist and founder of tamilnet.org, Taraki Sivaram, who was kidnapped in a white van and found brutally murdered.

It should also be noted that disappearances and kidnapping of dissenters has long been a practice of successive governments in Sri Lanka; this history includes the treatment of JVP rebel insurgencies in 1971 and the 1980s. Sri Lanka has one of the world's highest numbers of enforced disappearances, between 60,000 and 100,000 since the late 1980s (Amnesty International 2017). In 2018, former Sri Lanka Army general Sarath Fonseka claimed that former defence secretary Gotabaya Rajapaksa ordered wartime abductions to silence his administration's critics (Ambrose and Yeo 2018).

12. Scientists are not immune to being punished for the inaccuracies of their predictions. In Italy, scientists were arrested after the L'Aquila earthquake in March of 2009 for failing to warn the public. See Cohen (2012).
13. On the readjustments of popular religious beliefs in Batticaloa, eastern Sri Lanka, see Lawrence (2010). On tsunami recovery in southern India, see Hastrup (2011).

3. ENDURANCE

1. I use *endurance* rather than *resilience*, which is recognized as a term that depoliticizes the ways in which people rebuild their lives in situations of adversity. See Cons (2018); Duffield (2012); Grove (2013). On resilience and cultures of preparedness, see Neocleous (2013).
2. See Stefanie Graeter's (2017) ethnographic work in Peru on "life beyond bare life."
3. I am also inspired here by Srinivas's work on "utopia," where for her, the "sense of utopia . . . is not the will to power or the attempt to control and fix temporality and spatiality" (2015, 159) but rather improvisation and possibility in the face of precariousness.
4. "Build Back Better" was a phrase and also charge that was undertaken by international aid organizations and governments in the reconstruction phase after the tsunami. It was made famous by former US president Bill Clinton, who was a UN Special Envoy to the Tsunami: "The only way to redeem that kind of loss is to empower and dignify those people who have suffered. That is 'building back better'—but it also means letting them define it" ("Lessons Learned from Tsunami Recovery: Key Propositions for Building Back Better," Office of the UN Secretary-General, Special Envoy to the Tsunami, December 2006). See Seale-Feldman (2020) on "Building Back Otherwise" after the 2015 earthquake in Nepal. For a critical examination of "Building Back Better" after the Indian Ocean Tsunami in Southern India, see Swamy (2021).
5. Health officials were initially concerned about outbreaks of waterborne diseases such as cholera and malaria, but the salt water inundated

freshwater mosquito breeding grounds, while the distribution of bottled water helped to thwart a cholera outbreak. Instead, the ocean water mixed with sand, silt, and bacteria caused secondary infections among survivors. These conditions included "tsunami lung," a "necrotising pneumonia," and prolonged infections of cuts and wounds sustained by survivors of the tsunami.

6. See Ruwanpura and Humphries (2004) on how war-affected women maintain households as "mundane heroines" in eastern Sri Lanka. As Ruwanpura (2006) also stresses, the hardships experienced by women in eastern Sri Lanka cannot be singularly attributed to the war but must be understood in the context of more complicated social dynamics and institutional factors.

7. See Gutkowski (2018) for a discussion of the "theft" of time as a colonial, ethno-nationalist technology of power in occupied Palestine. See also Kotef and Amir (2011) and Abourahme (2011). On "waiting" as a colonial project, see Chakrabarty (2000).

8. Though the voices are taken directly from my notes, real names have been changed.

9. Following Rajchman (2007), new sensibilities beyond singular linear conceptualizations of time (and history) for Deleuze included "juxtapositions" as opposed to "successions."

10. According to the Ministry of Telecommunication and Information Technology in Sri Lanka, by 2009, the island had a nearly 80 percent mobile phone saturation rate. Mobile phones are cheaper options than personal computers to gain access to the internet. Moreover, for a small fee, many can sign up to receive up-to-date news messages and broadcasts from news outlets in the country. See also Whitaker (2004) on diasporic Tamil nationalism and the internet. For an analysis of the use of text messaging in popular movements, see Vicente Rafael (2003).

11. *Grama sevaka* was the name of the village level officer until it was changed to what is now G.N., or *grama niladhari*.

12. Since 2008, Sri Lanka's crumbling clock towers have been rebuilt and digitized.

4. REITERATION

1. Initially the British East India Company took over Dutch territory, but the area was transferred to the Crown and became a crown colony in 1802. Ceylon's first survey department was created in 1800 (Barrow 2008).

2. As Thongchai writes, "The term geo-body is mine. But the definition of the term is neither strict nor inclusive. Readers will find that it is flexible

enough to convey meanings concerning the territoriality of the nation" (1994, 17). A "geo-body" of nation is meant to suggest how the nation is an effect of or produced through technologies such as maps. The geo-body is the "component of the life of a nation. It is a source of pride, loyalty, love, passion, bias, hatred, reason, unreason. It generates many other conceptions and practices about nationhood" (1994, 17).

3. There have been waves of critical approaches to GIS. Early critiques were informed by positivism (see Pickles 1995), while later approaches of "critical GIS" (see Crampton 2011; Propen 2006) were informed by feminist critiques (see Schuurman and Pratt 2002). Feminist geography (see, for example, Elwood 2006; Kwan 2002; Schuurman 2000) has been instrumental in delineating the knowledge politics of maps and the ways in which social inequalities and relations are reproduced in them.

4. James Scott characterizes maps as artifacts of the rationality of state power and renders them "fictions of reality" or "fictitious facts on paper" (Scott 1998, 83). This critique stems from the notion that maps are not the territory and do not represent ground reality. Hence, state efforts to realize organized schemes of cities, communities, and development projects tend to fail, because state power imposes its agency through rational, simplified, abstract terms that are unable to deal with the complexities of reality. On these grounds, state power could be said to "fail" because its vision as articulated in the map is not real or does not capture reality on the ground "accurately." While I do not contest that the map may not represent "reality" as such, such a critique leaves untroubled the logic of rationality itself. To challenge state power because of a discrepancy between its rational technologies and "reality" is one that remains inside the logic of rational state power.

5. On the significance of information and information distribution and control after disasters, see Finn (2018); see also Ananny and Finn (2020). For an intriguing examination on the complexities of "fixing" images, see Grant (2022).

6. Geographic Information Systems (GIS) emerged from a military-industrial era and became especially salient during the Cold War. As Caren Kaplan (2006) points out, computer science and satellite programs grew as the United States attempted to militarize space and develop weapons that could be used for defending against or attacking other superpowers. The technologies used in satellite imaging and mapping stemmed from a military-industrial complex. GPS, for example, was originally a military technology; the first satellites were launched by the US Department of Defense in the 1970s. Satellite imagery and satellite maps are perhaps most powerful for their sense of pictorial realism (see Dumit 2004 on digital imaging; Pickles 2012 on GIS). Indeed, as critical

geographers point out, digital mapping could be seen as taking transparency to its extreme, with the availability of information real-time mapping, Google maps, satellites—or, perhaps more pointedly put, a "libidinal investment in an ontology of transparency and depth: investing subjects, bodies and spaces 'in depth'" (Doel 2006, 343). Technology carries a fetishized truth-making force; hence, suspicion of digital mapping technologies is not necessarily unwarranted. This seduction of transparency and objectivity can lead to a too-sanitized, not-quite-real version of reality (Feldman 1994) where the ground truth is not properly represented (Cosgrove 2001). Critical geographers have labored to historicize and trace the development of these technologies, which are now common, if not ubiquitous.

7. See Burns (2014a, 2014b) and Shringarpure (2018) on the promises and pitfalls of "digital humanitarianism."
8. The posters were commissioned by the National Movement Against Terrorism, a JHU (Jathika Hela Urumaya, "National Sinhala Heritage," or "National Heritage Party")-affiliated organization. The JHU is a Sinhala nationalist political party.
9. Here I am reminded of Sumathi Ramaswamy's work on *tamilparru* or what she calls "Tamil devotion." Tamil language devotion evokes and works itself into a national imagination, although, as she claims, it is not subsumed by nationalism. The "passions" for Tamil language can only be partially contained within the metanarrative of nationalism or a singular conception of the nation (Ramaswamy 1997). Certainly, one might consider Prabhakaran as iconic, as the figure for one form of Tamil nationalism in Sri Lanka, but the Tamil movement cannot be completely subsumed by him.
10. On how photography and photographic circulations and reframings in Jaffna work to contest state narratives of the war in northern Sri Lanka while also articulating the political possibilities of a Tamil nation, see Buthpitiya (2023).
11. Indeed, as Crampton and Krygier (2005, 12) write, "if the map is a specific set of power-knowledge claims, then not only the state but others could make competing and equally powerful claims."

POSTSCRIPT

1. On China's involvement in the war, see Popham (2010); see also Macan-Markar (2020).
2. For a helpful summary of anti-Muslim violence, see Haniffa (2021).
3. For a more detailed outline of the existing legal structures and pathways to manage the pandemic, see Centre for Policy Alternatives (2020).

References

AAAS. 2009. "High-Resolution Satellite Imagery and the Conflict in Sri Lanka." Accessed December 11, 2023. https://www.aaas.org/resources/geotech/high-resolution-satellite-imagery-and-conflict-sri-lanka.

Abeydeera, Ananda. 1993. "Mapping as a Vital Element of Administration in the Dutch Colonial Government of Maritime Sri Lanka, 1658–1796." *Imago Mundi* 45:101–11.

Abourahme, Nasser. 2011. "Spatial Collisions and Discordant Temporalities: Everyday Life between Camp and Checkpoint." *International Journal of Urban and Regional Research* 35 (2): 453–61.

ACAPS. 2015. "ACAPS Briefing Note: Sri Lanka—Floods." January 7, 2015. https://reliefweb.int/report/sri-lanka/acaps-briefing-note-sri-lanka-floods-7-january-2015.

Adams, Vincanne. 2013. *Markets of Sorrow, Labors of Faith: New Orleans in the Wake of Katrina*. Durham, NC: Duke University Press.

Adams, Vincanne, Michelle Murphy, and Adele Clarke. 2009. "Anticipation: Technoscience, Life, Affect, Temporality." *Subjectivity* 28 (1): 246–65.

Adey, Peter, Ben Anderson, and Stephen Graham. 2015. "Introduction: Governing Emergencies: Beyond Exceptionality." *Theory, Culture and Society* 32 (2): 3–17.

Agamben, Giorgio. 1999. "Absolute Immanence." In *Potentialities*, edited by Daniel Heller-Roazen, 220–39. Palo Alto, CA: Stanford University Press.

Ahmann, Chloe. 2018. "'It's Exhausting to Create an Event out of Nothing': Slow Violence and the Manipulation of Time." *Cultural Anthropology* 33 (1): 142–71.

Alatout, Samer. 2008. "'States' of Scarcity: Water, Space, and Identity Politics in Israel, 1948–59." *Environment and Planning D: Society and Space* 26 (6): 959–82.

Aljazeera. 2023. "Sri Lanka to Slash Military by a Third to Cut Costs." January 13, 2023. https://www.aljazeera.com/news/2023/1/13/bankrupt-sri-lanka-to-slash-military-by-third-to-cut-costs.

Allen, Lori. 2008. "Getting by the Occupation: How Violence Became Normal during the Second Palestinian Intifada." *Cultural Anthropology* 23 (3): 453–87.

Allison, Anne. 2013. *Precarious Japan*. Durham, NC: Duke University Press.

Alonso, Ana María. 1994. "The Politics of Space, Time and Substance: State Formation, Nationalism, and Ethnicity." *Annual Review of Anthropology* 23 (1): 379–405.

Alphonsus, Daniel. 2021. "Sri Lanka's Post-war Defence Budget: Overspending and Underprotection." *South Asia Scan* 15 (November). https://www.isas.nus.edu.sg/papers/sri-lankas-post-war-defence-budget-overspending-and-underprotection/.

Amarasuriya, Harini, and Jonathan Spencer. 2015. "'With That, Discipline Will Also Come to Them': The Politics of the Urban Poor in Postwar Colombo." *Current Anthropology* 56 (s11): s66–s75.

Ambrose, Drew, and Sarah Yeo. 2018. "Abduction and Forced Disappearance: Sri Lanka's Missing Thousands." *Aljazeera*, May 14, 2008. https://www.aljazeera.com/features/2018/5/14/abduction-and-forced-disappearance-sri-lankas-missing-thousands.

Amnesty International. 2017. "Sri Lanka—Victims of Disappearance Cannot Wait Any Longer for Justice." April 3, 2017. https://www.amnesty.org/en/latest/news/2017/04/sri-lanka-victims-of-disappearance-cannot-wait-any-longer-for-justice-2/.

Amoore, Louise. 2013. *The Politics of Possibility: Risk and Security beyond Probability*. Durham, NC: Duke University Press.

Ananny, Mike, and Megan Finn. 2020. "Anticipatory News Infrastructures: Seeing Journalism's Expectations of Future Publics in Its Sociotechnical Systems." *New Media and Society* 22 (9): 1600–1618.

Anderson, Ben. 2010. "Preemption, Precaution, Preparedness: Anticipatory Action and Future Geographies." *Progress in Human Geography* 34 (6): 777–98.

Anderson, Ben, and Peter Adey. 2012. "Anticipating Emergencies: Technologies of Preparedness and the Matter of Security." *Security Dialogue* 43 (2): 99–117.

Anderson, Benedict. 1991. *Imagined Communities: Reflections on the Origin and Spread of Nationalism*. Rev. ed. New York: Verso.

Aneez, Shihar, and Aditi Shah. 2019. "Hardline Sri Lanka Monk Calls for Buddhist Sinhalese Government." *Reuters*, July 7, 2019. https://www.reuters.com/article/us-sri-lanka-buddhist/hardline-sri-lanka-monk-calls-for-buddhist-sinhalese-government-idUSKCN1U2078.

Aretxaga, Begoña. 1999. *Shattering Silence: Women, Nationalism, and Political Subjectivity in Northern Ireland*. Princeton, NJ: Princeton University Press.

Aretxaga, Begoña. 2001. "The Sexual Games of the Body Politic: Fantasy and State Violence in Northern Ireland." *Culture, Medicine, and Psychiatry* 25:1–27.

Aretxaga, Begoña. 2003. "Maddening States." *Annual Review of Anthropology* 32:393–410.

Arif, Yasmeen. 2016. *Life, Emergent: The Social in the Afterlives of Violence*. Minneapolis: University of Minnesota Press.

Arulingam, Swasthika, Marisa De Silva, and Farzana Haniffa. 2022. "The People Revolt: Sri Lanka Panel." Panel discussion, University of California Santa Cruz Center for South Asian Studies, October 11, 2022. https://youtu.be/9VAK2PaOaZM?feature=shared.

Azoulay, Ariella. 2008. *The Civil Contract of Photography*. Translated by Rela Mazali and Ruvik Danieli. New York: Zone Books.

Banerjee, Dwaipayan. 2020. *Enduring Cancer: Life, Death, and Diagnosis in Delhi*. Durham, NC: Duke University Press.

Barrios, Roberto E. 2016. "What Does Catastrophe Reveal for Whom? The Anthropology of Crises and Disasters at the Onset of the Anthropocene." *Annual Review of Anthropology* 46 (1): 151–66.

Barrow, Ian. 2008. *Surveying and Mapping in Colonial Sri Lanka: 1800–1900*. Colombo: Vijitha Yapa.

Bass, Daniel. 2008. "Paper Tigers on the Prowl: Rumors, Violence and Agency in the Up-Country of Sri Lanka." *Anthropological Quarterly* 81 (1): 269–95.

Bastian, Sunil. 1999. "The Failure of State Formation, Identity Conflict, and Civil Society Responses: The Case of Sri Lanka." CCR Working Papers, no. 2. Centre for Conflict Resolution, Department of Peace Studies, University of Bradford, Bradford, UK.

Bear, Laura. 2016. "Time as Technique." *Annual Review of Anthropology* 45 (1): 487–502.

Beck, Ulrich. 1992. *Risk Society: Towards a New Modernity*. London: Sage.

Beckett, Greg. 2020. *There Is No More Haiti: Between Life and Death in Port-au-Prince*. Berkeley: University of California Press.

Benadusi, Mara. 2020. "Blurred Memories: War and Disaster in a Buddhist Sinhala Village." *Focaal*, no. 88, 89–102.

Benjamin, Walter. (1968) 1986. *Illuminations*. Edited by Hannah Arendt. New York: Schocken.

Benjamin, Walter. 1999. *The Arcades Project*. Translated by Howard Eiland and Kevin McLaughlin. Cambridge, MA: Belknap Press.

Benton, Adia. 2017. "Whose Security? Militarization and Securitization during West Africa's Ebola Outbreak." In *The Politics of Fear: Médecins* sans Frontières and the West African Ebola Epidemic, edited by Michiel Hofman and Sokhieng Au, 25–50. New York: Oxford University Press.

Bergson, Henri. (1896) 2004. *Matter and Memory*. Translated by Nancy Margaret Paul and W. Scott Palmer. New York: Dover.

Berlant, Lauren. 1994. *The Anatomy of National Fantasy: Hawthorne, Utopia, and Everyday Life*. Chicago: University of Chicago Press.

Berlant, Lauren. 2011. *Cruel Optimism*. Durham, NC: Duke University Press.

Berry, Maya J., Claudia Chávez Argüelles, Shanya Cordis, Sarah Ihmoud, and Elizabeth Velásquez Estrada. 2017. "Toward a Fugitive Anthropology: Gender, Race, and Violence in the Field." *Cultural Anthropology* 32 (4): 537–65.

Bhan, Mona. 2014. *Counterinsurgency, Democracy, and the Politics of Identity in India: From Warfare to Welfare?* New York: Routledge.

Bhan, Mona, and Purnima Bose. 2020. "Canine Counterinsurgency in Indian-Occupied Kashmir." *Critique of Anthropology* 40 (3): 341–63.

Bier, Jess. 2017. *Mapping Israel, Mapping Palestine: How Occupied Landscapes Shape Scientific Knowledge*. Cambridge, MA: MIT Press.

Blanchot, Maurice. 1995. *The Writing of the Disaster*. Translated by Ann Smock. Lincoln: University of Nebraska Press.

Bloch, Ernst. 1995. *The Principle of Hope*. Vol. 1. Cambridge, MA: MIT Press.

Bond, David. 2013. "Governing Disaster: The Political Life of the Environment during the BP Oil Spill." *Cultural Anthropology* 28 (4): 694–715.

Bonilla, Yarimar. 2020. "The Coloniality of Disaster: Race, Empire, and the Temporal Logics of Emergency in Puerto Rico, USA." *Political Geography* 78:1–42.

Bhungalia, Lisa. 2023. *Elastic Empire: Refashioning War through Aid in Palestine*. Stanford, CA: Stanford University Press.

Burns, Ryan. 2014a. "Moments of Closure in the Knowledge Politics of Digital Humanitarianism." *Geoforum* 53:51–62.

Burns, Ryan. 2014b. "Rethinking Big Data in Digital Humanitarianism: Practices, Epistemologies, and Social Relations." *GeoJournal* 80 (4): 477–90.

Buthpitiya, Vindhya. 2023. "'The Truth Is in the Soil': The Political Work of Photography in Northern Sri Lanka." In *Citizens of Photography: The Camera and the Political Imagination*, edited by Christopher Pinney, PhotoDemos Collective, Naluwembe Binaisa, Vindhya Buthpitiya, Konstantinos Kalantzis, Ileana L. Selejan, and Sokphea Young, 63–110. Durham, NC: Duke University Press.

Butler, Judith. 2004. *Precarious Life: The Powers of Mourning and Violence*. New York: Verso.

Button, Gregory, and Mark Schuller, eds. 2016. *Contextualizing Disaster*. New York: Berghahn.

Buultjens, J. W., I. Ratnayake, and W. K. Athula Chammika Gnanapala. 2016. "Post-conflict Tourism Development in Sri Lanka: Implications for Building Resilience." *Current Issues in Tourism* 19 (4): 355–72.

Calhoun, Craig. 2017. "The Rhetoric of Nationalism." In *Everyday Nationhood*, edited by Michael Skey and Marco Antonsich, 17–30. London: Palgrave Macmillan.

Callon, Michel. 1986. "Some Elements of a Sociology of Translation: Domestication of the Scallops and the Fishermen of St Brieuc Bay." In *Power, Action and Belief: A New Sociology of Knowledge?*, edited by John Law, 196–233. London: Routledge.

Canadian Broadcasting Corporation. 2009. "Body of Tamil Tiger Leader Found, Sri Lanka's Army Says." May 19, 2009. https://www.cbc.ca/news/world/body-of-tamil-tiger-leader-found-sri-lanka-s-army-says-1.819412.

Carter, Rebecca Louise. 2019. *Prayers for the People: Homicide and Humanity in the Crescent City*. Chicago: University of Chicago Press.

Centre for Policy Alternatives. 2013. "The Need to Repeal and Replace the Prevention of Terrorism Act." Updated May 9, 2013. https://www.cpalanka.org/the-need-to-repeal-and-replace-the-prevention-of-terrorism-act-pta/.

Centre for Policy Alternatives. 2020. "Structures to Deal with COVID-19 in Sri Lanka: A Brief Comment on the Presidential Task Force." April 2020. https://www.cpalanka.org/wp-content/uploads/2020/04/FINAL-Presidential-Task-Force-on-COVID19-April-2020-copy.pdf.

Centre for Policy Alternatives. 2023. "A Brief Analysis of the *Aragalaya*." Colombo: Centre for Policy Alternatives—Social Indicator Report.

Centre for Policy Alternatives. 2024. "CPA Challenges the Anti-terrorism Bill (SC/SD 04/2024)." January 19, 2024. https://www.cpalanka.org/cpa-challenges-the-anti-terrorism-bill-sc-sd-04-2024/.

Chakrabarty, Dipesh. 2000. *Provincializing Europe: Postcolonial Thought and Historical Difference*. Princeton, NJ: Princeton University Press.

Chatterjee, Partha. 1993. *The Nation and Its Fragments: Colonial and Postcolonial Histories*. Princeton, NJ: Princeton University Press.

Chatterjee, Partha. 2004. *The Politics of the Governed: Reflections on Popular Politics in Most of the World*. New York: Columbia University Press.

Chatterjee, Partha. 2005. "The Nation in Heterogeneous Time." *Futures* 37 (9): 925–42.

Choi, Vivian. 2009. "A Safer Sri Lanka? Technology, Security and Preparedness in Post-tsunami Sri Lanka." In *Tsunami in a Time of War: Aid, Activism and Reconstruction in Sri Lanka and Aceh*, edited by Malathi de Alwis and Eva-Lotta Hedman, 191–216. Colombo: International Centre for Ethnic Studies.

Choi, Vivian. 2015. "Anticipatory States: Tsunami, War, and Insecurity in Sri Lanka." *Cultural Anthropology* 30 (2): 286–309.

Choi, Vivian. 2020. "Woman, Non-native, Other." *Commoning Ethnography* 3 (1): 123–28.

Choi, Vivian. 2021. "Infrastructures of Feeling: The Sense and Governance of Disasters in Sri Lanka." In *Disastrous Times: Beyond Environmental Crisis in Urbanizing Asia*, edited by Eli Elinoff and Tyson Vaughan, 83–101. Philadelphia: University of Pennsylvania Press.

Choy, Timothy, and Jerry Zee. 2015. "Condition—Suspension." *Cultural Anthropology* 30 (2): 210–23.

Clarke, Lee. 2001. *Mission Improbable: Using Fantasy Documents to Tame Disaster*. Chicago: University of Chicago Press.

Cohen, Joel. 2012. "A Seismic Crime." United Nations Office for Disaster Risk Reduction, March 6, 2012. https://www.undrr.org/news/seismic-crime.

Colombo Correspondent. 2019. "President Sirisena and PM Wickremesinghe Take Opposite Stances on Lanka War Crime Allegations." *South Asia Journal*, March 8, 2019. http://southasiajournal.net/president-sirisena-and-pm-wickremesinghe-take-opposite-stances-on-lanka-war-crime-allegations/.

Cons, Jason. 2018. "Staging Climate Security: Resilience and Heterodystopia in the Bangladesh Borderlands." *Cultural Anthropology* 33 (2): 266–94.

Coomaraswamy, Radhika, and Charmaine de los Reyes. 2004. "Rule by Emergency: Sri Lanka's Postcolonial Constitutional Experience." *International Journal of Constitutional Law* 2 (2): 272–95.

Cooper, Melinda. 2006. "Pre-empting Emergence: The Biological Turn in the War on Terror." *Theory, Culture and Society* 23 (4): 113–35.

Coronil, Fernando, and Julie Skurski, eds. 2005. *States of Violence*. Ann Arbor: University of Michigan Press.

Cosgrove, Denis. 2001. *Apollo's Eye: A Cartographic Genealogy of the Earth in the Western Imagination*. Baltimore, MD: Johns Hopkins University Press.

Craib, Raymond. 2004. *Cartographic Mexico: A History of State Fixations and Fugitive Landscapes*. Durham, NC: Duke University Press.

Crampton, Jeremy W. 2011. *Mapping: A Critical Introduction to Cartography and GIS*. Hoboken, NJ: Wiley-Blackwell.

Crampton, Jeremy, and Krygier, John. 2005. "An Introduction to Critical Cartography." *ACME: An International Journal for Critical Geographies* 4 (1): 11–33.

Daniel, E. Valentine. 1996. *Charred Lullabies: Chapters in an Anthropography of Violence*. Princeton, NJ: Princeton University Press.

Darusman, Marzuki, Steven Ratner, and Yasmin Sooka. 2011. "Report of the Secretary-General's Panel of Experts on Accountability in Sri Lanka."

UN Security Council Report. https://www.securitycouncilreport.org/un-documents/document/poc-rep-on-account-in-sri-lanka.php.

Das, Veena. 2007. *Life and Words: Violence and the Descent into the Ordinary*. Berkeley: University of California Press.

Das, Veena, Arthur Kleinman, Margaret Lock, Mamphela Ramphele, and Pamela Reynolds. 2001. *Remaking a World: Violence, Social Suffering, and Recovery*. Berkeley: University of California Press.

Das, Veena, and Deborah Poole, eds. 2004. *Anthropology in the Margins of the State*. Santa Fe: SAR Press.

Daston, Lorraine, and Peter Galison. 2007. *Objectivity*. Princeton, NJ: Zone.

D'Avella, Nicholas. 2019. *Concrete Dreams: Practice, Value, and Built Environments in Post-crisis Buenos Aires*. Durham, NC: Duke University Press.

Davis, Christina. 2020. *The Struggle for a Multilingual Future: Youth and Education in Sri Lanka*. Oxford: Oxford University Press.

Dawdy, Shannon. 2006. "The Taphonomy of Disaster and the (Re)Formation of New Orleans." *American Anthropologist* 108 (4): 719–30.

de Alwis, Malathi. 1995. "Gender, Politics and the 'Respectable Lady.'" In *Unmaking the Nation: The Politics of Identity and History in Modern Sri Lanka*, edited by Pradeep Jeganathan and Qadri Ismail, 137–57. Colombo: Social Scientists' Association.

de Alwis, Malathi. 2004. "The 'Purity' of Displacement and the Reterritorialization of Longing: Muslim IDPs in Northwestern Sri Lanka." In *Sites of Violence: Gender and Conflict Zones*, edited by Wenona Giles and Jennifer Hyndman, 213–31. Berkeley: University of California Press.

de Alwis, Malathi. 2010. "Sri Lanka Must Respect Memory of War." *Guardian*, May 4, 2010. https://www.theguardian.com/commentisfree/2010/may/04/sri-lanka-must-respect-war-memory.

de Alwis, Malathi, and Eva-Lotta Hedman, eds. 2009. *Tsunami in a Time of War: Aid, Activism and Reconstruction in Sri Lanka and Aceh*. Colombo: International Centre for Ethnic Studies.

Debord, Guy. 1994. *The Society of the Spectacle*. Translated by Donald Nicholson-Smith. New York: Zone.

de la Cadena, Marisol. 2011. "Indigenous Cosmopolitics in the Andes: Conceptual Reflections beyond 'Politics.'" *Cultural Anthropology* 25 (2): 334–70.

Deleuze, Gilles. 1989. *Cinema 2: The Time-Image*. Translated by Hugh Tomlinson and Robert Galeta. Minneapolis: University of Minnesota Press.

Deleuze, Gilles. 1992. "Postscript on the Societies of Control." *October* 59 (Winter): 3–7.

Deleuze, Gilles. 1997. "Immanence: A Life . . ." *Theory, Culture and Society* 14 (2): 3–7.

de Mel, Neloufer. 2007. *Militarizing Sri Lanka: Popular Culture, Memory and Narrative in the Armed Conflict*. New York: Sage.

Department of Census and Statistics. 2012. "A Brief Analysis of Population and Housing Characteristics." Colombo, Sri Lanka.

de Silva, Amarasiri M. W. 2009. "Ethnicity, Politics and Inequality: Post-tsunami Humanitarian Aid Delivery in Ampara District, Sri Lanka." *Disasters* 33 (2): 253–73.

De Silva, Ranjith Premalal. 2009. "Interpretation Not Substantiated without Ground Verification." *Daily News*, May 6, 2009. http://archives.dailynews.lk/2009/05/06/fea02.asp.

Dillard, Annie. 1982. *Teaching a Stone to Talk: Expeditions and Encounters*. New York: HarperPerennial.

Dillon, Michael, and Luis Lobo-Guerrero. 2009. "The Biopolitical Imaginary of Species-Being." *Theory, Culture and Society* 26 (1): 1–23.

Disaster Management Centre. 2012. "Annual Report—2012." Colombo: Disaster Management Centre.

Doel, Marcus. 2006. "The Obscenity of Mapping." *Area* 38 (3): 343–45.

Downey, Gary Lee, and Joseph Dumit. 1997. "Locating and Intervening: An Introduction." In *Cyborgs and Citadels: Anthropological Interventions in Emerging Sciences and Technologies*, edited by Gary Lee Downey and Joseph Dumit, 5–29. Santa Fe: SAR Press.

Dowty, Rachel, and Barbara Allen, eds. 2010. *Dynamics of Disaster: Lessons on Risk, Response and Recovery*. London: Earthscan.

Duara, Prasenjit. 1996. *Rescuing History from the Nation: Questioning Narratives of Modern China*. Chicago: University of Chicago Press.

Dudden, Alexis. 2012. "The Ongoing Disaster." *Journal of Asian Studies* 71 (2): 345–59.

Duffield, Mark. 2012. "Challenging Environments: Danger, Resilience and the Aid Industry." *Security Dialogue* 43 (5): 475–92.

Dumit, Joseph. 1999. "Objective Brains, Prejudicial Images." *Science in Context* 12 (1): 173–201.

Dumit, Joseph. 2004. *Picturing Personhood: Brain Scans and Biomedical Identity*. Princeton, NJ: Princeton University Press.

Dumit, Joseph. 2012. *Drugs for Life: How Pharmaceutical Companies Define Our Health*. Durham, NC: Duke University Press.

Dumit, Joseph. 2014. "Writing the Implosion: Teaching the World One Thing at a Time." *Cultural Anthropology* 29 (2): 344–62.

Eder, Jens, and Charlotte Klonk, eds. 2017. *Image Operations: Visual Media and Political Conflict*. Manchester: Manchester University Press.

Edney, Matthew H. 1997. *Mapping an Empire: The Geographical Construction of British India, 1765–1843*. Chicago: University of Chicago Press.

Elinoff, Eli, and Tyson Vaughan, eds. 2021. *Disastrous Times: Beyond Environmental Crisis in Urbanizing Asia*. Philadelphia: University of Pennsylvania Press.

Elwood, Sarah. 2006. "Beyond Cooptation or Resistance: Urban Spatial Politics, Community Organizations, and GIS-Based Spatial Narratives." *Annals of the Association of American Geographers* 96 (2): 323–41.

Elwood, Sarah, and Agnieszka Leszczynski. 2013. "New Spatial Media, New Knowledge Politics." *Transactions of the Institute of British Geographers* 38 (4): 544–59.

Enia, Jason. 2008. "Peace in Its Wake? The 2004 Tsunami and Internal Conflict in Indonesia and Sri Lanka." *Journal of Public and International Affairs* 19 (1): 7–27.

ESRI. "What Is GIS?" Accessed August 8, 2024. https://www.esri.com/en-us/what-is-gis/overview.

Farmer, Paul. 2011. *Haiti after the Earthquake*. New York: Public Affairs.

Fassin, Didier, and Mariella Pandolfi. 2010. *Contemporary States of Emergency: The Politics of Military and Humanitarian Interventions*. New York: Zone.

Fedricks, Krishantha, Farzana Haniffa, Anushka Kahandagamage, Chulani Kodikara, Kaushalya Kumarasinghe, and Jonathan Spencer. 2023. "Snapshots from the Struggle, Sri Lanka April–May 2022." *Anthropology Now* 14 (1–2): 21–38.

Feldman, Allen. 1994. "On Cultural Anesthesia: From Desert Storm to Rodney King." *American Ethnologist* 21 (2): 404–18.

Feldman, Allen. 2015. *Archives of the Insensible: Of War, Photopolitics, and Dead Memory*. Chicago: University of Chicago Press.

Fernando, Jude Lal. 2013. "War by Other Means: Expansion of Siṃhala Buddhism into the Tamil Region in 'Post-war' Īlam." In *Buddhism among Tamils in Tamilakam and Īlam*, edited by Peter Schalk, 174–239. Uppsala, Sweden: Uppsala University.

Fernando, Jude Lal. 2015. "Lost Lives and a Missed Political Opportunity: The Politics of Conflict and Peace in Post-tsunami Sri Lanka." *Asian Journal of Peacebuilding* 3 (2): 137–63.

Fernando, Nathasha. 2020. "Strategic Demilitarization in Sri Lanka: Paradoxes and Trajectories." JIA SIPA, December 9, 2020. https://jia.sipa.columbia.edu/news/strategic-demilitarization-sri-lanka-paradoxes-and-trajectories.

Finn, Megan. 2018. *Documenting Aftermath: Information Infrastructures in the Wake of Disasters*. Cambridge, MA: MIT Press.

Fisch, Michael. 2022. "Japan's Extreme Infrastructure: Fortress-ification, Resilience, and Extreme Nature." *Social Science Japan Journal* 25 (2): 331–52. https://doi.org/10.1093/ssjj/jyac011.

Fischer, Michael M. J. 2003. *Emergent Forms of Life and the Anthropological Voice*. Durham, NC: Duke University Press.

Fonseka, Bhavani. 2020. "Uncomfortable Truths with the Pandemic Response in Sri Lanka." *Groundviews*, May 11, 2020. https://groundviews

.org/2020/11/05/uncomfortable-truths-with-the-pandemic-response-in-sri-lanka/.

Fonseka, Bhavani, and Uvin Dissanayake. 2021. "Sri Lanka's Vistas of Prosperity and Splendour: A Critique of Promises Made and Present Trends." Colombo: Centre for Policy Alternatives.

Fonseka, Bhavani, Luwie Ganeshathasan, and Asanga Welikala. 2021. "Sri Lanka: Pandemic-Catalyzed Democratic Backsliding." In *Covid-19 in Asia: Law and Policy Contexts*, edited by Victor V. Ramraj. Online ed. New York: Oxford Academic.

Fonseka, Bhavani, and Mirak Raheem. 2010. "Land in the Eastern Province—Politics, Policy and Conflict." Centre for Policy Alternatives. https://www.cpalanka.org/land-in-the-eastern-provincepolitics-policy-and-conflict/.

Fonseka, Bhavani, and Kushmila Ranasinghe. 2021. "Sri Lanka's Accelerated Democratic Decay amidst a Pandemic." In *Is the Cure Worse Than the Disease? Reflections on COVID Governance in Sri Lanka*, edited by Pradeep Peiris, 29–60. Colombo: Centre for Policy Alternatives.

Fortun, Kim. 2001. *Advocacy after Bhopal: Environmentalism, Disaster, New Global Orders*. Chicago: University of Chicago Press.

Fortun, Kim. 2012. "Ethnography in Late Industrialism." *Cultural Anthropology* 27 (3): 446–64.

Fortun, Kim, Scott Gabriel Knowles, Vivian Choi, Paul Jobin, Miwao Matsumoto, Pedro de la Torre III, Max Liboiron, and Luis Felipe R. Murillo. 2017. "Researching Disaster from an STS Perspective." In *The Handbook of Science and Technology Studies*, 4th ed., edited by Ulrike Felt, Rayvon Fouché, Clark A. Miller, and Laurel Smith-Doerr, 1003–28. Cambridge, MA: MIT Press.

Fortun, Mike. 2006. *Promising Genomics: Iceland and deCODE Genetics in a World of Speculation*. Berkeley: University of California Press.

Foucault, Michel. 1984. "Of Other Spaces: Utopias and Heterotopias." Translated by Jay Miskowiec. https://web.mit.edu/allanmc/www/foucault1.pdf. Orig. pub. *Architecture/Mouvement/Continuité*, October 1984.

Foucault, Michel. 2003. *"Society Must Be Defended": Lectures at the Collège de France 1975–1976*. Translated by David Macey. New York: Picador.

Foucault, Michel. 2007. *Security, Territory, Population: Lectures at the Collège de France 1977–1978*. Translated by Graham Burchell. New York: Macmillan.

Freeman, Elizabeth. 2010. *Time Binds: Queer Temporalities, Queer Histories*. Durham, NC: Duke University Press.

Funahashi, Daena. 2016. "Bringing Medical Anthropology Back into the Fold: An Interview with Daena Funahashi." *Fieldsights*, April 15, 2016.

https://culanth.org/fieldsights/bringing-medical-anthropology-back-into-the-fold-an-interview-with-daena-funahashi.

Gajaweera, Nalika. 2015. "Buddhist Cosmopolitan Ethics and Transnational Secular Humanitarianism in Sri Lanka." In *Religion and the Politics of Development*, edited by Philip Fountain, Robin Bush, and R. Michael Feener, 105–28. London: Palgrave Macmillan.

Gamburd, Michele. 2013. *The Golden Wave: Culture and Politics after Sri Lanka's Tsunami Disaster*. Bloomington: Indiana University Press.

Government of Sri Lanka. 2005. *Post Tsunami Recovery and Reconstruction: Progress, Challenges, Way Forward*. Colombo: Government of Sri Lanka.

Govindrajan, Radhika. 2018. *Animal Intimacies: Interspecies Relatedness in India's Central Himalayas*. Chicago: University of Chicago Press.

Graeter, Stefanie. 2017. "To Revive an Abundant Life: Catholic Science and Neoextractivist Politics in Peru's Mantaro Valley." *Cultural Anthropology* 32 (1): 117–48.

Grant, Jenna. 2022. *Fixing the Image: Ultrasound and the Visuality of Care in Phnom Penh*. Seattle: University of Washington Press.

Greenhouse, Carol J. 1996. *A Moment's Notice: Time Politics across Cultures*. Ithaca, NY: Cornell University Press.

Grosz, Elizabeth. 2013. "Habit Today: Ravaisson, Bergson, Deleuze and Us." *Body and Society* 19 (2–3): 217–39.

Groundviews. 2012. "Horrible Rise of Disappearances in Post-war Sri Lanka Continues Unabated." April 5, 2012. http://groundviews.org/2012/04/05/horrible-rise-of-disappearances-in-post-war-sri-lanka-continues-unabated.

Grove, Kevin. 2013. "Hidden Transcripts of Resilience: Power and Politics in Jamaican Disaster Management." *Resilience: International Policies, Practices, and Discourses* 1 (3): 193–209.

Guggenheim, Michael. 2014. "Introduction: Disasters as Politics—Politics as Disasters." Supplement, *Sociological Review* 62 (S1): 1–16.

Gunawardene, Nalaka. 2011. "DON'T PANIC! Predicting Earthquakes or Triggering Mass Hysteria?" *Groundviews*, April 19, 2011. https://groundviews.org/2011/04/19/don't-panic-predicting-earthquakes-or-triggering-mass-hysteria/.

Gunewardena, Nandini. 2010. "Peddling Paradise, Rebuilding Serendib: The 100-Meter Refugees versus the Tourism Industry in Post-tsunami Sri Lanka." In *Capitalizing on Catastrophe: Neoliberal Strategies in Disaster Reconstruction*, edited by Nandini Gunewardena and Mark Schuller, 69–92. Lanham, MD: AltaMira.

Gusterson, Hugh. 2004. *People of the Bomb: Portraits of America's Nuclear Complex*. Minneapolis: University of Minnesota Press.

Gutkowski, Natalia. 2018. "Governing through Timescape: Israeli Sustainable Agriculture Policy and the Palestinian-Arab Citizens." *International Journal of Middle East Studies* 50 (3): 471–92.

Hall, Stuart. 1996. "When Was the 'Post-colonial'? Thinking at the Limit." In *The Post-colonial Question: Common Skies, Divided Horizons*, edited by Iain Chambers and Lidia Curti, 242–60. New York: Routledge.

Hall, Stuart. 2011. "The Neoliberal Revolution: Thatcher, Blair, Cameron—the Long March of Neoliberalism Continues." *Soundings* 48:9–28.

Haniffa, Farzana. 2008. "Piety as Politics amongst Muslim Women in Contemporary Sri Lanka." *Modern Asian Studies* 42 (2–3): 347–75. https://doi.org/10.1017/S0026749X07003137.

Haniffa, Farzana. 2015a. "Competing for Victim Status: Northern Muslims and the Ironies of Sri Lanka's Post-war Transition." *Stability: International Journal of Security and Development* 4 (1): 1–18.

Haniffa, Farzana. 2015b. "Fecund Mullas and Goni Billas: Gendered Nature of Anti Muslim Rhetoric in Post-war Sri Lanka." *South Asianist Journal* 4 (1). https://www.southasianist.ed.ac.uk/article/view/1308.

Haniffa, Farzana. 2021. "What Is Behind the Anti-Muslim Measures in Sri Lanka?" *aljazeera*, April 12, 2021. https://www.aljazeera.com/opinions/2021/4/12/what-is-behind-the-anti-muslim-measures-in-sri-lanka.

Hansen, Thomas Blom, and Finn Stepputat. 2006. "Sovereignty Revisited." *Annual Review of Anthropology* 35:295–315.

Haraway, Donna. 1988. "Situated Knowledges: The Science Question in Feminism and the Privilege of Partial Perspective." *Feminist Studies* 14 (3): 575–99.

Haraway, Donna. 2016. *Staying with the Trouble: Making Kin in the Chthulucene*. Durham, NC: Duke University Press.

Harley, John Brian. 1988. *Maps, Knowledge, and Power*. Cambridge: Cambridge University Press.

Harley, John Brian. 2001. *The New Nature of Maps: Essays in the History of Cartography*. Baltimore, MD: Johns Hopkins University Press.

Hartman, Saidiya V. 2002. "The Time of Slavery." *South Atlantic Quarterly* 101 (4): 757–77.

Harwell, Emily. 2000. "Remote Sensibilities: Discourses of Technology and the Making of Indonesia's Natural Disaster." *Development and Change* 31 (1): 307–40.

Hasbullah, Shahul. 2001. *Muslim Refugees: The Forgotten People in Sri Lanka's Ethnic Conflict*. Nuraicholai: Research and Action Forum for Social Development.

Hasbullah, Shahul. 2016. "Northern Muslim Expulsion and Tamil Leadership." *Colombo Telegraph*, November 2, 2016. https://www.colombotelegraph.com/index.php/northern-muslim-expulsion-tamil-leadership/.

Hasbullah, Shahul, and Benedikt Korf. 2009. "Muslim Geographies and the Politics of Purification in Sri Lanka after the 2004 Tsunami." *Singapore Journal of Tropical Geography* 30 (2): 248–64.

Hasbullah, Shahul, and Benedikt Korf. 2013. "Muslim Geographies, Violence and the Antinomies of Community in Eastern Sri Lanka." *Geographical Journal* 179 (1): 32–43.

Hastrup, Frida. 2011. *Weathering the World: Recovery in the Wake of the Tsunami in a Tamil Fishing Village*. New York: Berghahn.

Heath-Kelly, Charlotte. 2018. "Forgetting ISIS: Enmity, Drive and Repetition in Security Discourse." *Critical Studies on Security* 6 (1): 85–99.

Helmreich, Stefan. 2006. "Time and the Tsunami." *Reconstruction* 6 (3). http://reconstruction.digitalodu.com/Issues/063/helmreich.shtml.

Hewage, Thushara. 2014. "Ideology, Ethnicity, and the Critique of Postconflict in Sri Lanka." Hot Spots, *Fieldsights*, March 24, 2014. https://culanth.org/fieldsights/ideology-ethnicity-and-the-critique-of-postconflict-in-sri-lanka.

Hewamanne, Sandya. 2013. "The War Zone in My Heart: The Occupation of Southern Sri Lanka." In *Everyday Occupations: Experiencing Militarism in South Asia and the Middle East*, edited by Kamala Visweswaran, 60–84. Philadelphia: University of Pennsylvania Press.

Hewitt, Kenneth. 2021. "Afterword. 'Acts of Men': Disasters Neglected, Preventable, and Moral." In *Critical Disaster Studies*, edited by Jacob A. C. Remes and Andy Horowitz, 184–92. Philadelphia: University of Pennsylvania Press.

Hoffman, Susannah M., and Anthony Oliver-Smith, eds. 2002. *Catastrophe and Culture: The Anthropology of Disaster*. Santa Fe: SAR Press.

Holbraad, Martin, and Morten Axel Pedersen, eds. 2013. *Times of Security: Ethnographies of Fear, Protest and the Future*. New York: Routledge.

Holt, John, ed. 2016. *Buddhist Extremists and Muslim Minorities: Religious Conflict in Contemporary Sri Lanka*. New York: Oxford University Press.

Honig, Bonnie. 2009. *Emergency Politics: Paradox, Law, Democracy*. Princeton, NJ: Princeton University Press.

Hoole, Rajan, Daya Somasundaram, Kopalasingham Sritharan, and Rajani Thiranagama. 1990. *The Broken Palmyra: The Tamil Crisis in Sri Lanka: An Inside Account*. Claremont, CA: Sri Lanka Studies Institute.

HuffPost. 2009. "Tamil Tiger Leader Vellupillai Prabhakaran Survival Conspiracy Theories Emerging." June 21, 2009. https://www.huffpost.com/entry/tamil-tiger-leader-vellup_n_206531.

Human Rights Watch. 2018a. "Locked Up without Evidence: Abuses under Sri Lanka's Prevention of Terrorism Act." January 29, 2018. https://www.hrw.org/report/2018/01/29/locked-without-evidence/abuses-under-sri-lankas-prevention-terrorism-act#.

Human Rights Watch. 2018b. "Sri Lanka Draft Counter Terrorism Act of 2018: Human Rights Watch Submission to Parliament." October 21,

2018. https://www.hrw.org/news/2018/10/21/sri-lanka-draft-counter-terrorism-act-2018.

Hyndman, Jennifer. 2007. "The Securitization of Fear in Post-tsunami Sri Lanka." *Annals of the Association of American Geographers* 97 (2): 361–72.

Hyndman, Jennifer, and Amarnath Amarasingam. 2014. "Touring 'Terrorism': Landscapes of Memory in Post-war Sri Lanka." *Geography Compass* 8 (8): 560–75.

Hyndman, Jennifer, and Malathi de Alwis. 2004. "Bodies, Shrines, and Roads: Violence, (Im)mobility and Displacement in Sri Lanka." *Gender, Place and Culture* 11 (4): 535–57.

India Today. 2009. "Exclusive: Encounter of Prabhakaran." May 19, 2009. https://www.indiatoday.in/world/photo/exclusive-encounter-of-prabhakaran-361944-2009-05-19/1.

International Commission of Jurists. 2012. "Authority without Accountability: The Crisis of Impunity in Sri Lanka." Geneva, Switzerland. https://www.icj.org/wp-content/uploads/2013/01/ICJ-Srilanka-Report.pdf.

International Commission of Jurists. 2021a. "Sri Lanka: 'De-radicalization' Regulations Should Be Immediately Withdrawn." March 18, 2021. https://www.icj.org/sri-lank-de-radicalization-regulations-should-be-immediately-withdrawn/.

International Commission of Jurists. 2021b. "Sri Lanka: New Anti-terror Regulations Aimed at Organizations Further Undermine the Rule of Law." April 15, 2021. https://www.icj.org/sri-lanka-new-anti-terror-regulations-aimed-at-organizations-further-undermine-the-rule-of-law/.

Ismail, S. M. M., M. A. M. Rameez, and M. M. Fazil. 2005. "Muslim Perspective from the East." In *Dealing with Diversity: Sri Lankan Discourses on Peace and Conflict*, edited by Georg Frerks and Bart Klem, 151–60. The Hague: Netherlands Institute of International Relations.

Ives, Sarah. 2019. "'More-Than-Human' and Less-Than-Human': Race, Botany, and the Challenge of Multispecies Ethnography." *Catalyst: Feminism, Theory, Technoscience* 5 (2): 2–5.

Jameson, Fredric. 2002. "The Dialectics of Disaster." *South Atlantic Quarterly* 101 (2): 297–304.

Jayawardena, Kumari. 2003. "Ethnicity and Sinhala Consciousness." In *July '83 and After*, edited by N. Rajasingham, 47–86. Colombo: International Centre for Ethnic Studies.

Jayawardena, Kumari, and Malathi de Alwis. 1996. *Embodied Violence: Communalising Female Sexuality in South Asia*. London: Zed.

Jazeel, Tariq. 2013. *Sacred Modernity: Nature, Environment, and the Postcolonial Geographies of Sri Lankan Nationhood*. Oxford: Oxford University Press.

Jeganathan, Pradeep. 2000. "On the Anticipation of Violence: Modernity and Identity in Southern Sri Lanka." In *Anthropology, Development and Modernities: Exploring Discourses, Counter-tendencies and Violence*, edited by Alberto Arce and Norman Long, 111–25. London: Routledge.

Jeganathan, Pradeep, and Qadri Ismail, eds. 1995. *Unmaking the Nation: The Politics of Identity and History in Modern Sri Lanka*. Colombo: Social Scientists' Association.

Jegathesan, Mythri. 2013. "Bargaining in a Labor Regime: Plantation Life and the Politics of Development in Sri Lanka." PhD diss., Columbia University.

Jegathesan, Mythri. 2015. "Deficient Realities: Expertise and Uncertainty among Tea Plantation Workers in Sri Lanka." *Dialectical Anthropology* 39 (3): 255–72.

Jegathesan, Mythri. 2019. *Tea and Solidarity: Tamil Women and Work in Postwar Sri Lanka*. Seattle: University of Washington Press.

Jegathesan, Mythri. 2020. "A Body Writing." *Commoning Ethnography* 3 (1): 129–35.

Jessop, Bob. 2005. "Gramsci as a Spatial Theorist." *Critical Review of International Social and Political Philosophy* 8 (4): 421–37.

Junaid, Mohamad. 2013. "Death and Life under Occupation: Space, Violence, and Memory in Kashmir." In *Everyday Occupations: Experiencing Militarism in South Asia and the Middle East*, edited by Kamala Visweswaran, 158–90. Philadelphia: University of Pennsylvania Press.

Kadirgamar, Ahilan. 2013. "The Question of Militarisation in Post-war Sri Lanka." *Economic and Political Weekly* 48 (7): 42–46.

Kadirgamar, Ahilan. 2020. "Polarization, Civil War, and Persistent Majoritarianism in Sri Lanka." In *Political Polarization in South and Southeast Asia: Old Divisions, New Dangers*, 53–66. Washington, DC: Carnegie Endowment for Peace.

Kadirgamar, Ahilan. 2022. "Rethinking Sri Lanka's Economic Crisis." *Himal South Asian*, February 22, 2022. https://www.himalmag.com/rethinking-sri-lankas-economic-crisis-interview-ahilan-kadirgamar-2022/.

Kanapathipillai, Valli. 2009. *Citizenship and Statelessness in Sri Lanka: The Case of the Tamil Estate Workers*. New York: Anthem.

Kaplan, Caren. 2006. "Precision Targets: GPS and the Militarization of US Consumer Identity." *American Quarterly* 58 (3): 693–714.

Keck, Frédéric. 2020. *Avian Reservoirs: Virus Hunters and Birdwatchers in Chinese Sentinel Posts*. Durham, NC: Duke University Press.

Keenan, Alan. 2010. "Building the Conflict Back Better: The Politics of Tsunami Relief and Reconstruction in Sri Lanka." In *Tsunami Recovery in Sri Lanka: Ethnic and Regional Dimensions*, edited by Dennis McGilvray and Michele Gamburd. New York: Routledge.

Kelly, John. 1998. "Time and the Global: Against the Homogeneous, Empty Communities in Contemporary Social Theory." *Development and Change* 29 (4): 839–71.

Khalili, Laleh. 2010. "The Location of Palestine in Global Counterinsurgencies." *International Journal of Middle East Studies* 42 (3): 413–33.

Kim, Gloria. 2016. "Pathogenic Nation-Making: Media Ecologies and American Nationhood under the Shadow of Viral Emergence." *Configurations* 24 (4): 441–70.

Kimura, Aya. 2016. *Radiation Brain Moms and Citizen Scientists: The Gender Politics of Food Contamination after Fukushima*. Durham, NC: Duke University Press.

Klein, Naomi. 2005. *The Shock Doctrine: The Rise of Disaster Capitalism*. New York: Picador.

Klem, Bart. 2018. "The Problem of Peace and the Meaning of 'Post-war.'" *Conflict, Security and Development* 18 (3): 233–55.

Klima, Alan. 2002. *The Funeral Casino: Meditation, Massacre, and Exchange with the Dead in Thailand*. Princeton, NJ: Princeton University Press.

Knowles, Scott. 2013. *The Disaster Experts: Mastering Risk in Modern America*. Philadelphia: University of Pennsylvania Press.

Knowles, Scott. 2020. "Slow Disaster in the Anthropocene: A Historian Witnesses Climate Change on the Korean Peninsula." *Daedalus* 149 (4): 192–206.

Knowles, Scott, and Zachary Loeb. 2021. "The Voyage of the *Paragon*: Disaster as Method." In *Critical Disaster Studies*, edited by Jacob A. C. Remes and Andy Horowitz, 11–31. Philadelphia: University of Pennsylvania Press.

Korf, Benedikt. 2006a. "Commentary on the Special Section on the Indian Ocean Tsunami: Disasters, Generosity and the Other." *Geographical Journal* 172 (3): 245–47.

Korf, Benedikt. 2006b. "Who Is the Rogue? Discourse, Power and Spatial Politics in Post-war Sri Lanka." *Political Geography* 25 (3): 279–97. https://doi.org/10.1016/j.polgeo.2005.12.007.

Korf, Benedikt, Shahul Habullah, Pia Hollenbach, and Bart Klem. 2010. "The Gift of Disaster: The Commodification of Good Intentions in Post-tsunami Sri Lanka." *Disasters* 34 (s1): s60–s77.

Kotef, Hagar, and Merav Amir. 2011. "Between Imaginary Lines: Violence and Its Justifications at the Military Checkpoints in Occupied Palestine." *Theory, Culture and Society* 28 (1): 55–80.

Krishna, Sankaran. 1994. "Cartographic Anxiety: Mapping the Body Politic in India." *Alternatives: Global, Local, Political* 19 (4): 507–21.

Krishna, Sankaran. 1996. "Producing Sri Lanka: J. R. Jayewardene and Postcolonial Identity." *Alternatives: Global, Local, Political* 21 (3): 303–20.

Krishna, Sankaran. 1999. *Postcolonial Insecurities: India, Sri Lanka, and the Question of Nationhood*. Minneapolis: University of Minnesota Press.

Kwan, Mei-Po. 2002. "Feminist Visualization: Re-envisioning GIS as a Method in Feminist Geographic Research." *Annals of the Association of American Geographers* 92 (4): 645–61.

Lakoff, Andrew. 2008. "The Generic Biothreat, or, How We Became Unprepared." *Cultural Anthropology* 23 (3): 399–428.

Lakoff, Andrew, and Stephen J. Collier. 2015. "Vital Systems Security: Reflexive Biopolitics and the Government of Emergency." *Theory, Culture and Society* 32 (2): 19–51.

Lal, Vinay. 1994. "Anti-terrorist Legislation: A Comparative Study of India, the United Kingdom, and Sri Lanka." *Lokayan Bulletin* 11, no. 1 (July/August): 5–24.

Latour, Bruno. 1986. "Visualization and Cognition: Thinking with Eyes and Hands." *Knowledge and Society: Studies in the Sociology of Culture Past and Present* 6:1–40.

Latour, Bruno. 1993. *We Have Never Been Modern*. Cambridge, MA: Harvard University Press.

Lawrence, Patricia. 1997. "Work of Oracles, Silence of Terror: Notes on the Injury of War in Eastern Sri Lanka." PhD diss., University of Colorado.

Lawrence, Patricia. 2010. "The Sea Goddess and the Fishermen: Religion and Recovery in Navalady, Sri Lanka." In *Tsunami Recovery in Sri Lanka: Ethnic and Regional Dimensions*, edited by Dennis B. McGilvray and Michele R. Gamburd. New York: Routledge.

Lay, Thorne, Hiroo Kanamori, Charles J. Ammon, Meredith Nettles, Steven N. Ward, Richard C. Aster, Susan L. Beck, et al. 2005. "The Great Sumatra-Andaman Earthquake of 26 December 2004." *Science* 308 (5725): 1127–33.

Le Billon, Philippe, and Arno Waizenegger. 2007. "Peace in the Wake of Disaster? Secessionist Conflicts and the 2004 Indian Ocean Tsunami." *Transactions of the Institute of British Geographers* 32 (3): 411–27.

Liboiron, Max, and David Wachsmuth. 2013. "The Fantasy of Disaster Response: Governance and Social Action during Hurricane Sandy." *Social Text Online*, October 29, 2013. https://socialtextjournal.org/periscope_article/the-fantasy-of-disaster-response-governance-and-social-action-during-hurricane-sandy/.

Lloréns, Hilda. 2021. *Making Livable Worlds: Afro-Puerto Rican Women Building Environmental Justice*. Seattle: University of Washington Press.

Lomnitz, Claudio. 2001. *Deep Mexico, Silent Mexico: An Anthropology of Nationalism*. Minneapolis: University of Minnesota Press.

Lowe, Celia. 2010. "Viral Clouds: Becoming H5N1 in Indonesia." *Cultural Anthropology* 25 (4): 625–49.

Lutz, Catherine. 2002. "Making War at Home in the United States: Militarization and the Current Crisis." *American Anthropologist* 104 (3): 723–35.

Lynch, Caitrin. 2007. *Juki Girls, Good Girls: Gender and Cultural Politics in Sri Lanka's Global Garment Industry*. Ithaca, NY: Cornell University Press.

Macan-Markar, Marwaan. "Sri Lanka Assured of China's Help in Burying Post-war Obligations." *NikkeiAsia*, October 28, 2020. https://asia.nikkei.com/Politics/International-relations/Sri-Lanka-assured-of-China-s-help-in-burying-post-war-obligations.

Mahadev, Neena. 2014. "Conversion and Anti-conversion in Contemporary Sri Lanka: Pentecostal Christian Evangelism and Theravada Buddhist Views on the Ethics of Religious Attraction." In *Proselytizing and the Limits of Religious Pluralism in Contemporary Asia*, edited by Juliana Finucane and R. Michael Feener, 211–35. Singapore: ARI–Springer.

Mandelbaum, Moran M. 2019. *The Nation/State Fantasy: A Psychoanalytical Genealogy of Nationalism*. Cham, Switzerland: Springer Nature.

Manor, James, ed. 1984. *Sri Lanka in Change and Crisis*. New York: St. Martin's.

Martin, Emily. 1990. "Toward an Anthropology of Immunology: The Body as Nation State." *Medical Anthropology Quarterly* 4 (4): 410–26.

Masco, Joseph. 2008. "'Survival Is Your Business': Engineering Ruins and Affect in Nuclear America." *Cultural Anthropology* 23 (2): 361–98.

Masco, Joseph. 2014. *The Theater of Operations: National Security Affect from the Cold War to the War on Terror*. Durham, NC: Duke University Press.

Massumi, Brian. 2005. "Fear (The Spectrum Said)." *positions: East Asia Cultures Critique* 13 (1): 31–48.

Maunaguru, Sidharthan. 2019. *Marrying for a Future: Transnational Sri Lankan Tamil Marriages in the Shadow of War*. Seattle: University of Washington Press.

Mavhunga, Clapperton Chakanetsa. 2011. "Vermin Beings: On Pestiferous Animals and Human Game." *Social Text* 29 (1): 151–76.

McGilvray, Dennis B. 2001. *Tamil and Muslim Identities in the East*. Colombo: Marga Institute.

McGilvray, Dennis B. 2008. *Crucible of Conflict: Tamil and Muslim Society on the East Coast of Sri Lanka*. Durham, NC: Duke University Press.

McGilvray, Dennis B., and Michele Ruth Gamburd. 2010. *Tsunami Recovery in Sri Lanka: Ethnic and Regional Dimensions*. London: Routledge.

McGilvray, Dennis B., and Mirak Raheem. 2007. *Muslim Perspectives on the Sri Lankan Conflict*. Washington, DC: East-West Center Washington.

Ministry of Defence. 2009. "When the Camera Lies for Terror." May. https://www.army.lk/si/node/37768.

Ministry of Disaster Management. 2005. Disaster Management Act No. 13. Colombo: Ministry of Disaster Management.

Ministry of Environmental and Natural Resources. 2007. *Sri Lanka Strategy for Sustainable Development*. Colombo: Ministry of Environmental and Natural Resources.

Mitchell, Timothy. 2002. *Rule of Experts: Egypt, Techno-politics, Modernity*. Berkeley: University of California Press.

Mitchell, W. J. T. 2004. *What Do Pictures Want? The Lives and Loves of Images*. Chicago: University of Chicago Press.

Moinudeen, Sakina. 2021. "Ethno-centric Pandemic Governance: The Muslim Community in Sri Lanka's COVID Response." In *Is the Cure Worse Than the Disease? Reflections on COVID Governance in Sri Lanka*, edited by Pradeep Peiris, 111–28. Colombo: Centre for Policy Alternatives.

Mol, Annemarie. 2002. *The Body Multiple: Ontology in Medical Practice*. Durham, NC: Duke University Press.

Moore, Mick. 1985. *The State and Peasant Politics in Sri Lanka*. Cambridge: Cambridge University Press.

Moore, Mick. 1993. "Thoroughly Modern Revolutionaries: The JVP in Sri Lanka." *Modern Asian Studies* 27 (3): 593–642.

Morimoto, Ryo. 2012. "Shaking Grounds, Unearthing Palimpsests: Semiotic Anthropology of Disaster." *Semiotica*, no. 192, 263–74.

Morimoto, Ryo. 2023. *Nuclear Ghost: Atomic Livelihoods in Fukushima's Gray Zone*. Berkeley: University of California Press.

Moten, Fred. 2003. *In the Break: The Aesthetics of the Black Radical Tradition*. Minneapolis: University of Minnesota Press.

Murphy, Michelle. 2017. "Alterlife and Decolonial Chemical Relations." *Cultural Anthropology* 32 (4): 494–503. https://doi.org/10.14506/ca32.4.02.

National Science Foundation. 2005. "Analysis of the Sumatra-Andaman Earthquake Reveals Longest Fault Rupture Ever." May 19, 2005. https://www.nsf.gov/news/news_summ.jsp?cntn_id=104179.

Navaro-Yashin, Yael. 2002. *Faces of the State: Secularism and Public Life in Turkey*. Princeton, NJ: Princeton University Press.

Navaro-Yashin, Yael. 2012. *The Make-Believe Space: Affective Geography in a Postwar Polity*. Durham, NC: Duke University Press.

Navarro, Tami, Bianca Williams, and Attiya Ahmad. 2013. "Sitting at the Kitchen Table: Fieldnotes from Women of Color in Anthropology." *Cultural Anthropology* 28 (3): 443–63.

Nelson, Diane. 2009. *Reckoning: The Ends of War in Guatemala*. Durham, NC: Duke University Press.

Nelson, Diane, and Carlota McAllister. 2013. *War by Other Means: Aftermath in Post-genocide Guatemala*. Durham, NC: Duke University Press.

Neocleous, Mark. 2013. "Resisting Resilience." *Radical Philosophy* 178 (March–April). https://www.radicalphilosophy.com/commentary/resisting-resilience.

Nesiah, Vasuki, and Alan Keenan. 2004. "Human Rights and Sacred Cows: Framing Violence, Disappearing Struggles." In *From the Margins of Globalization: Critical Perspectives on Human Rights*, edited by Neve Gordan, 261–95. Lanham, MD: Lexington Books.

news.lk: The Government Official News Portal. 2014. "Sri Lanka Army Assists Flood Victims." December 22, 2014. https://news.lk/news/business/item/5268-sri-lanka-army-assists-flood-victims.

New York Times. 2010. "The 31 Places to Go in 2010." January 7, 2010. https://archive.nytimes.com/www.nytimes.com/2010/01/10/travel/10places.html.

Nixon, Rob. 2013. *Slow Violence and the Environmentalism of the Poor*. Cambridge, MA: Harvard University Press.

Nizamdeen, Farhan. 2023. "57 out of 77 Tsunami Early Warning Towers Out of Service—DMC." *Ceylon Today*, December 26, 2023.

Office of Foreign Disaster Assistance. 1979. "Disaster Case Report: Sri Lanka—Cyclone." Washington, DC: Agency for International Development.

Oliver-Smith, Anthony. 2002. "Theorizing Disasters." In *Catastrophe and Culture: The Anthropology of Disaster*, edited by Anthony Oliver-Smith and Susanna Hoffman, 23–47. Santa Fe: SAR Press.

Ong, Aihwa, and Stephen J. Collier, eds. 2005. *Global Assemblages: Technology, Politics, and Ethics as Anthropological Problems*. Malden, MA: Blackwell.

Ophir, Adi. 2010. "The Politics of Catastrophization: Emergency and Exception." In *Contemporary States of Emergency: The Politics of Military and Humanitarian Interventions*, edited by Didier Fassin and Mariella Pandolfi, 59–88. New York: Zone Books.

Orr, Jackie. 2006. *Panic Diaries: A Geneaology of Panic Disorder*. Durham, NC: Duke University Press.

Paprocki, Kasia. 2019. "All That Is Solid Melts into the Bay: Anticipatory Ruination and Climate Change Adaptation." *Antipode* 51 (1): 295–315. https://doi.org/10.1111/anti.12421.

Park, Jeffrey, Teh-Ru Alex Song, Jeroen Tromp, Emile Okal, Seth Stein, Genevieve Roult, Eric Clevede, et al. 2005. "Earth's Free Oscillations Excited by the 26 December 2004 Sumatra-Andaman Earthquake." *Science* 308 (5725): 1139–44.

Peebles, Patrick. 1990. "Colonization and Ethnic Conflict in the Dry Zone of Sri Lanka." *Journal of Asian Studies* 49 (1): 30–55. https://doi.org/10.2307/2058432.

Peebles, Patrick. 2001. *The Plantation Tamils of Ceylon*. London: Leicester University Press.

Peiris, Pradeep, ed. 2021. *Is the Cure Worse Than the Disease? Reflections on COVID Governance in Sri Lanka*. Colombo: Social Indicator, Centre for Policy Alternatives.

Perera, Amantha. 2005. "The Buffer Zone Fiasco." *Sunday Leader*, December 25, 2005.

Perry, Ronald W. 2007. "What Is a Disaster?" In *Handbook of Disaster Research*, edited by Havidán Rodríguez, Enrico L. Quarantelli, and Russell R. Dynes, 1–15. New York: Springer.

Peters, Laura E. R., and Ilan Kelman. 2020. "Critiquing and Joining Intersections of Disaster, Conflict, and Peace Research." *International Journal of Disaster Risk Science* 11:555–67.

Petryna, Adriana. 2002. *Life Exposed: Biological Citizens after Chernobyl*. Princeton, NJ: Princeton University Press.

Pickles, John, ed. 1995. *The Ground Truth: The Social Implications of Geographic Information Systems*. New York: Guilford.

Pickles, John. 2012. *A History of Spaces: Cartographic Reason, Mapping and the Geo-coded World*. London: Routledge.

Pieris, Anoma. 2015. "Arterial Blockages: The Catastrophic Itineraries of the Sri Lankan Civil War." *National Identities* 17 (2): 195–215.

Pieris, Anoma. 2018. *Sovereignty, Space and Civil War in Sri Lanka: Porous Nation*. New York: Routledge.

Pinfari, Marco. 2019. *Terrorists as Monsters: The Unmanageable Other from the French Revolution to the Islamic State*. New York: Oxford University Press.

Pinney, Christopher. 2015. "Civil Contract of Photography in India." *Comparative Studies of South Asia, Africa and the Middle East* 35 (1): 21–34.

Popham, Peter. 2010. "How Beijing Won Sri Lanka's Civil War." *Independent*, May 22, 2010. https://www.independent.co.uk/news/world/asia/how-beijing-won-sri-lanka-s-civil-war-1980492.html.

Porter, Natalie. 2019. *Viral Economies: Bird Flu Experiments in Vietnam*. Chicago: University of Chicago Press.

Povinelli, Elizabeth. 2011. *Economies of Abandonment: Social Belonging and Endurance in Late Liberalism*. Durham, NC: Duke University Press.

President's Media Division. 2020. "Army Commander Shavendra Silva Heads National Operation Centre for Prevention of COVID-19 Outbreak." http://www.pmdnews.lk/armycommander-shavendra-silva-heads-national-operation-centerforprevention-of-covid-19-outbreak/ [link no longer active].

Pritchard, Sara. 2010. *Confluence: The Nature of Technology and the Remaking of the Rhône*. Cambridge, MA: Harvard University Press.

Propen, Amy. 2006. "Critical GIS: Toward a New Politics of Location." *ACME: An International E-Journal for Critical Geographies* 4 (1): 131–44.

Puar, Jasbir. 2007. *Terrorist Assemblages: Homonationalism in Queer Times*. Durham, NC: Duke University Press.

Pugliese, Joseph. 2013. *State Violence and the Execution of Law: Biopolitical Caesurae of Torture, Black Sites, Drones*. New York: Routledge.

Quarantelli, Enrico Louis. 1998. *What Is a Disaster? Perspectives on the Question.* London: Routledge.

Radcliffe, Sarah. 2009. "National Maps, Digitalisation and Neoliberal Cartographies: Transforming Nation-State Practices and Symbols in Postcolonial Ecuador." *Transactions of the Institute of British Geographers* 34:426–44.

Rafael, Vicente. 2003. "The Cell Phone and the Crowd: Messianic Politics in the Contemporary Philippines." *Public Culture* 15 (3): 399–425.

Rajapaksa, Mahinda. 2009. "Address at the Victory Day Parade and National Tribute to the Security Forces Following the Defeat of Terrorism, Galle Face Green, June 3, 2009." *Sunday Observer*, November 15, 2009.

Rajasingham-Senanayake, Darini. 1999. "Democracy, Discontent and the Making of Bi-polar Ethnic Conflict in Post-colonial Sri Lanka." In *Ethnic Futures: The State and Identity Politics in Asia*, edited by Joanna Pfaff-Czarnecka, Darini Rajasingham-Senanayake, Ashis Nandy, and Edmund Terence Gomez, 99–134. New Delhi: Sage.

Rajasingham-Senanayake, Darini. 2001. *Identity on the Borderline: Multicultural History in a Moment of Danger.* Colombo: Marga Institute.

Rajchman, John. 2007. "Deleuze's Time: How the Cinematic Changes Our Idea of Art." *Diálogos* 42 (90): 239–60.

Ralph, Laurence. 2014. *Renegade Dreams: Living through Injury in Gangland Chicago.* Chicago: University of Chicago Press.

Ramaswamy, Sumathi. 1997. *Passions of the Tongue: Language Devotion in Tamil India, 1891–1970.* Berkeley: University of California Press.

Ramaswamy, Sumathi. 2010. *The Goddess and the Nation: Mapping Mother India.* Durham, NC: Duke University Press.

Rambukwella, Harshana. 2018. *The Politics and Poetics of Authenticity: A Cultural Genealogy of Sinhala Nationalism.* London: UCL Press.

Rampton, David. 2011. "'Deeper Hegemony': The Politics of Sinhala Nationalist Authenticity and the Failures of Power-Sharing in Sri Lanka." *Commonwealth and Comparative Politics* 49 (2): 245–73.

Reddy, Elizabeth. 2020. "Crying 'Crying Wolf': How Misfires and Mexican Engineering Expertise Are Made Meaningful." *Ethnos* 85 (2): 335–50.

Reddy, Elizabeth. 2023. *¡Alerta! Engineering on Shaky Ground.* Cambridge, MA: MIT Press.

Remes, Jacob A. C., and Andy Horowitz, eds. 2021. *Critical Disaster Studies.* Philadelphia: University of Pennsylvania Press. https://doi.org/10.2307/j.ctv1f45qvg.

Rheinberger, Hans-Jörg. 1994. "Experimental Systems: Historiality, Narration, and Deconstruction." *Science in Context* 7 (1): 65–81.

Rigg, Jonathan, Carl Grundy-Warr, Lisa Law, and May Tan-Mullins. 2008. "Grounding a Natural Disaster: Thailand and the 2004 Tsunami." *Asia Pacific Viewpoint* 49 (2): 137–54.

Ring, Laura. 2006. *Zenana: Everyday Peace in a Karachi Apartment Building*. Bloomington: Indiana University Press.
Rogers, John. 1994. "Post-orientalism and the Interpretation of Premodern and Modern Political Identities: The Case of Sri Lanka." *Journal of Asian Studies* 53 (1): 10–13.
Rogers, John. 2004. "Early British Rule and Social Classification in Lanka." *Modern Asian Studies* 38 (3): 625–47.
Roitman, Janet. 2013. *Anti-crisis*. Durham, NC: Duke University Press.
Russell, Jane. 1982. *Communal Politics under the Donoughmore Constitution, 1931–1947*. Dehiwala: Tisara Prakasakayo.
Ruwanpura, Kanchana. 2006. "Conflict and Survival: Sinhala Female-Headship in Eastern Sri Lanka." *Asian Population Studies* 2 (2): 187–200.
Ruwanpura, Kanchana. 2009. "Putting Houses in Place: Rebuilding Communities in Post-tsunami Sri Lanka." *Disasters* 33 (3): 436–56.
Ruwanpura, Kanchana, and Janet Humphries. 2004. "Mundane Heroines: Conflict, Ethnicity, Gender and Female-Headship in Eastern Sri Lanka." *Feminist Economics* 10 (2): 173–205.
Samarasinghe, Mahinda. 2009. "Disaster Risk Management and Planning." Transcript of address to the Second Session of the Global Platform for Disaster Risk Reduction in Geneva, Switzerland. *Daily News*, June 22, 2009. http://archives.dailynews.lk/2009/06/22/fea01.asp.
Samuels, Annemarie. 2019. *After the Tsunami: Disaster Narratives and the Remaking of Everyday Life in Aceh*. Honolulu: University of Hawai'i Press.
Saravanamuttu, Paikiasothy. 2021. Foreword to *Is the Cure Worse Than the Disease? Reflections on COVID Governance in Sri Lanka*, edited by Pradeep Peiris, vii–viii. Colombo: Centre for Policy Alternatives.
Sarjoon, Athambawa, Mohammad Agus Yusoff, and Nordin Hussin. 2016. "Anti-Muslim Sentiments and Violence: A Major Threat to Ethnic Reconciliation and Ethnic Harmony in Post-war Sri Lanka." *Religions* 7 (10): 125. https://doi.org/10.3390/rel7100125.
Scheper-Hughes, Nancy, and Philippe I. Bourgois, eds. 2004. *Violence in War and Peace: An Anthology*. New York: Blackwell.
Schnell, Izhak, and Christine Leuenberger. 2014. "Mapping Genres and Geopolitics: The Case of Israel." *Transactions of the Institute of British Geographers* 39 (4): 518–31.
Schonthal, Benjamin. 2016. "Configurations of Buddhist Nationalism in Modern Sri Lanka." In *Buddhist Extremists and Muslim Minorities: Religious Conflict in Sri Lanka*, edited by John Holt, 97–118. Oxford: Oxford University Press.
Schuller, Mark. 2016. *Humanitarian Aftershocks in Haiti*. Rutgers: Rutgers University Press.

Schuller, Mark, and Julie K. Maldonado. 2016. "Disaster Capitalism." *Annals of Anthropological Practice* 40 (1): 61–72.

Schuurman, Nadine. 2000. "Trouble in the Heartland: GIS and Its Critics in the 1990s." *Progress in Human Geography* 24 (4): 569–90.

Schuurman, Nadine, and Geraldine Pratt. 2002. "Care of the Subject: Feminism and Critiques of GIS." *Gender, Place and Culture* 9 (3): 291–99.

Scott, David. 2004. *Conscripts of Modernity: The Tragedy of Colonial Enlightenment*. Durham, NC: Duke University Press.

Scott, James C. 1998. *Seeing Like a State: How Certain Schemes to Improve the Human Condition Have Failed*. New Haven, CT: Yale University Press.

Seale-Feldman, Aidan. 2020. "The Work of Disaster: Building Back Otherwise in Post-earthquake Nepal." *Cultural Anthropology* 35 (2): 237–63.

Senewiratne, Hiran H. 2009. "Met Dept Quashes Tsunami Rumour." *Daily News*, July 20, 2009. http://archives.dailynews.lk/2009/07/20/news12.asp.

Shalhoub-Kevorkian, Nadera. 2014. "Palestinian Children as Tools for 'Legalized' State Violence." *Borderlands* 13 (1): 1–24.

Sharma, Sarah. 2014. *In the Meantime: Temporality and Cultural Politics*. Durham, NC: Duke University Press.

Sharpe, Christina Elizabeth. 2016. *In the Wake: On Blackness and Being*. Durham, NC: Duke University Press.

Shastri, Amita. 1999. "Estate Tamils, the Ceylon Citizenship Act of 1948, and Sri Lanka Politics." *Contemporary South Asia* 8 (1): 65–86.

Shastri, Amita. 2009. "Ending Ethnic Civil War: The Peace Process in Sri Lanka." *Commonwealth and Comparative Politics* 47 (1): 76–99.

Shneiderman, Sara, Bina Khapunghang Limbu, Jeevan Baniya, Manoj Suji, Nabin Rawal, Prakash Chandra Subedi, and Cameron David Warner. 2023. "House, Household, and Home: Revisiting Anthropological and Policy Frameworks through Postearthquake Reconstruction Experiences in Nepal." *Current Anthropology* 64 (5): 498–527.

Shringarpure, Bhakti. 2018. "Africa and the Digital Savior Complex." *Journal of African Cultural Studies* 32 (2): 178–94.

Silva, Kalinga Tudor, and Shahul Hasbullah. 2019. "Sacred Sites, Humanitarian Assistance and the Politics of Land Grabbing in Eastern Sri Lanka: The Case of Deegavapi." *Sri Lanka Journal of Sociology* 1 (1): 62–86.

Silva, Kalinga Tudor. 2014. *Decolonisation, Development and Disease: A Social History of Malaria in Sri Lanka*. Hyderabad: Orient Black Swan.

Simpson, Edward. 2020. "Forgetfulness without Memory: Reconstruction, Landscape, and the Politics of the Everyday in Post-earthquake Gujarat, India." *Journal of the Royal Anthropological Institute* 26 (4): 786–804.

Simpson, Edward, and Stuart Corbridge. 2008. "The Geography of Things That May Become Memories: The 2001 Earthquake in Kachchh-Gujarat and the Politics of Rehabilitation in the Pre-memorial Era." *Annals of the Association of American Geographers* 96 (3): 566–85.

Simpson, Edward, and Malathi de Alwis. 2008. "Remembering Natural Disaster: Politics and Culture of Memorials in Gujarat and Sri Lanka." *Anthropology Today* 24 (4): 6–12.

Sivasundaram, Sujit. 2013. *Islanded: Britain, Sri Lanka, and the Bounds of an Indian Ocean Colony*. Chicago: University of Chicago Press.

Smith, Neil. 2006. "There Is No Such Thing as a Natural Disaster." *Items: Insights from the Social Sciences*, June 11, 2006. https://items.ssrc.org/understanding-katrina/theres-no-such-thing-as-a-natural-disaster/.

Somasundaram, Daya. 1998. *Scarred Minds: The Psychological Impact of War on Sri Lankan Tamils*. New Delhi: Sage.

Somasundaram, Daya. 2013. *Scarred Communities: Psychosocial Impact of Man-Made and Natural Disasters on Sri Lankan Society*. New Delhi: Sage.

Sontag, Susan. 1965. "The Imagination of Disaster." *Commentary*, October 1965, 42–48.

Sontag, Susan. 2003. *Regarding the Pain of Others*. New York: Picador.

Sparke, Matthew. 1998. "A Map That Roared and an Original Atlas: Canada, Cartography, and the Narration of Nation." *Annals of the Association of American Geographers* 88 (3): 463–95.

Spencer, Jonathan, ed. 1992. *Sri Lanka: History and the Roots of Conflict*. London: Routledge.

Spencer, Jonathan. 2002. "The Vanishing Elite: The Political and Cultural Work of Nationalist Revolution in Sri Lanka." In *Elite Cultures: Anthropological Perspectives*, edited by Cris Shore and Stephen Nugent, 91–109. London: Routledge.

Spencer, Jonathan. 2003. "A Nation 'Living in Different Places': Notes on the Impossible Work of Purification in Postcolonial Sri Lanka." *Contributions to Indian Sociology* 37 (1–2): 1–23.

Spencer, Jonathan. 2008. "A Nationalism without Politics? The Illiberal Consequences of Liberal Institutions in Sri Lanka." *Third World Quarterly* 29 (3): 611–29.

Spencer, Jonathan, Jonathan Goodhand, Shahul Hasbullah, Bart Klem, Benedikt Korf, and Kalinga Tudor Silva, eds. 2014. *Checkpoint, Temple, Church and Mosque: A Collaborative Ethnography of War and Peace*. London: Pluto.

Spyer, Patricia, and Mary Margaret Steedly, eds. 2013. *Images That Move*. Santa Fe: SAR Press.

Sri Lanka Army. n.d. "Army Goes All Out for Dengue Eradication." Accessed May 5, 2024. https://www.army.lk/news/army-goes-all-out-dengue-eradication.

Sri Lankan Parliament Select Committee on Natural Disasters. 2005. *Report.* https://www.parliament.lk/uploads/comreports/1476092317082517.pdf.

Srinivas, Smriti. 2015. *A Place for Utopia: Urban Designs from South Asia.* Seattle: University of Washington Press.

Steuter, Erin, and Deborah Wills. 2010. "'The Vermin Have Struck Again': Dehumanizing the Enemy in Post 9/11 Media Representations." *Media, War and Conflict* 3 (2): 152–67.

Stewart, Kathleen. 1996. *A Space on the Side of the Road: Cultural Poetics in an "Other" America.* Princeton, NJ: Princeton University Press.

Stewart, Kathleen. 2007. *Ordinary Affects.* Durham, NC: Duke University Press.

Stewart, Michelle. n.d. "Columbine Changed Everything: Preparedness and Anticipatory Sensibilities." Manuscript in possession of the author.

Stirrat, Jock. 2006. "Competitive Humanitarianism: Relief and the Tsunami in Sri Lanka." *Anthropology Today* 22 (5): 11–16.

Strassler, Karen. 2010. *Refracted Visions: Popular Photography and National Modernity in Java.* Durham, NC: Duke University Press.

Subramaniam, Banu. 2001. "The Aliens Have Landed! Reflections on the Rhetoric of Biological Invasions." *Meridians* 2 (1): 26–40.

Sunday Observer. 2009. "Dawn of a Great Future for Mother Lanka." November 15, 2009. https://archives.sundayobserver.lk/2009/11/15/Pres.asp?id=s03.

Sur, Malini. 2020. "Time at Its Margins: Cattle Smuggling across the India-Bangladesh Border." *Cultural Anthropology* 35 (4): 546–74.

Swamy, Raja. 2021. *Building Back Better in India: Development, NGOs, and Artisanal Fishers after the 2004 Tsunami.* Tuscaloosa: University of Alabama Press.

Tadiar, Neferti Xina M. 2004. *Fantasy Production: Sexual Economies and Other Philippine Consequences for the New World Order.* Hong Kong: Hong Kong University Press.

Tambiah, Stanley Jeyaraja. 1986. *Sri Lanka: Ethnic Fratricide and the Dismantling of Democracy.* Chicago: University of Chicago Press.

Tambiah, Stanley Jeyaraja. 1992. *Buddhism Betrayed? Religion, Politics, and Violence in Sri Lanka.* Chicago: University of Chicago Press.

Tambiah, Stanley Jeyaraja. 1997. *Leveling Crowds: Ethnonationalist Conflicts and Collective Violence in South Asia.* Berkeley: University of California Press.

Taussig, Michael T. 1992. *The Nervous System.* New York: Routledge.

Telford, John, and John Cosgrave. 2007. "The International Humanitarian System and the 2004 Indian Ocean Earthquake and Tsunamis." *Disasters* 31 (1): 1–28.

Tennekoon, Serena. 1983. "Newspaper Nationalism: Sinhala Identity as Historical Discourse." In *Sri Lanka: History and the Root of Conflict*, edited by Jonathan Spencer, 205–26. London: Routledge.

Tennekoon, Serena. 1988. "Rituals of Development: The Accelerated Mahaväli Development Program of Sri Lanka." *American Ethnologist* 15 (2): 294–310.

Thiranagama, Sharika. 2011. *In My Mother's House: Civil War in Sri Lanka*. Philadelphia: University of Pennsylvania Press.

Thiranagama, Sharika. 2020a. "Figurations of Menace." *Immanent Frame*, April 1, 2020. https://tif.ssrc.org/2020/04/01/figurations-of-menace/.

Thiranagama, Sharika. 2020b. "The Leader as Image: Prabhakaran and the Visual Regimes of the Liberation Tigers of Tamil Eelam." *Tasveer Ghar*, October 16, 2020. http://www.tasveergharindia.net/essay/prabhakaran-visual-ltte.

Thiranagama, Sharika. 2022. "Figures of Menace: Militarisation in Post-war Sri Lanka." *South Asia: Journal of South Asian Studies* 45 (1): 183–203.

Thongchai, Winichakul. 1994. *Siam Mapped: A History of the Geo-body of a Nation*. Manoa: University of Hawaiʻi Press.

Thottam, Jyoti. 2009. "The Man Who Tamed the Tamil Tigers." *Time*, July 13, 2009. https://time.com/archive/6947319/the-man-who-tamed-the-tamil-tigers/.

Tierney, Kathleen J., Michael K. Lindell, and Ronald W. Perry. 2001. *Facing the Unexpected: Disaster Preparedness and Response in the United States*. Washington, DC: Joseph Henry.

Tironi, Manuel. 2014. "Atmospheres of Indagation: Disasters and the Politics of Excessiveness." Supplement, *Sociological Review* 62 (S1): 114–34.

Tironi, Manuel, Israel Rodriguez-Giralt, and Michael Guggenheim, eds. 2014. *Disasters and Politics: Materials, Experiments, Preparedness*. New York: Wiley-Blackwell.

Tsing, Anna. 2005. *Friction: An Ethnography of Global Connection*. Princeton, NJ: Princeton University Press.

Tsing, Anna. 2011. "The News in the Provinces." In *Cultural Citizenship in Island Southeast Asia: Nation and Belonging in the Hinterlands*, edited by Renato Rosaldo, 192–220. Berkeley: University of California Press.

Turnbull, David. 2000. *Masons, Tricksters and Cartographers: Comparative Studies in the Sociology of Scientific and Indigenous Knowledge*. New York: Routledge.

Udagama, Deepika. 2015. "An Eager Embrace: Emergency Rule and Authoritarianism in Republican Sri Lanka." In *Reforming Sri Lankan Presidentialism: Provenance, Problems and Prospects*, edited by Asanga Welikala, 286–333. Colombo: Centre for Policy Alternatives.

UNDRR. n.d. "DRR and UNDRR's History." Accessed January 15, 2025. https://www.undrr.org/our-work/history.

UNISDR. 2007. "Hyogo Framework for Action 2005–2015. Building the Resilience of Nations and Communities to Disasters." Geneva, Switzerland: United Nations International Strategy for Disaster Risk Reduction.

UNISDR. 2019. "What Is Disaster Risk Reduction?" https://www.unisdr.org/who-we-are/what-is-drr.

UNOHCHR (United Nations Office of the High Commissioner for Human Rights). 2020. "COVID-19: States Should Not Abuse Emergency Measures to Suppress Human Rights—UN Experts." UN press release, March 16, 2020. https://www.ohchr.org/en/press-releases/2020/03/covid-19-states-should-not-abuse-emergency-measures-suppress-human-rights-un.

UTHR. 1995. "Information Bulletin No. 6: Report on the Situation in the Batticaloa and Amparai Districts." June 9, 1995. https://uthr.org/bulletins/bul6.htm.

UTHR. 2009. "A Marred Victory and a Defeat Pregnant with Foreboding." University Teachers for Human Rights, Jaffna, Sri Lanka. https://uthr.org/SpecialReports/spreport32.htm.

Uyangoda, Jayadeva. 2005. "Ethnic Conflict, the State and the Tsunami Disaster in Sri Lanka." *Inter-Asia Cultural Studies* 6 (3): 341–52.

Uyangoda, Jayadeva. 2008. "The Janatha Vimukthi Peramuna Split." *Economic and Political Weekly* 43 (18): 8–10.

Uyangoda, Jayadeva. 2010. "Introduction." In *Prevention of Terrorism Act (PTA): A Critical Analysis*, edited by Bertram Bastiampillai, Rohan Edirisinghe, and N. Kandasamy, 1–3. Colombo: Centre for Human Rights and Development.

Valluvan, Sivamohan. 2019. *The Clamour of Nationalism: Race and Nation in Twenty-First-Century Britain.* Manchester: Manchester University Press.

Vázquez-Arroyo, Antonio Y. 2013. "How Not to Learn from Catastrophe: Habermas, Critical Theory and the 'Catastrophization' of Political Life." *Political Theory* 41 (5): 738–65.

Venugopal, Rajesh. 2010. "Sectarian Socialism: The Politics of Sri Lanka's Janatha Vimukthi Peramuna (JVP)." *Modern Asian Studies* 44 (3): 567–602.

Venugopal, Rajesh. 2011. "The Politics of Market Reform at a Time of Civil War: Military Fiscalism in Sri Lanka." *Economic and Political Weekly* 46 (49): 67–75.

Venugopal, Rajesh. 2015. "Demonic Violence and Moral Panic in Postwar Sri Lanka: Explaining the Grease Devil Crisis." *Journal of Asian Studies* 74 (3): 615–37.

Venugopal, Rajesh. 2018. *Nationalism, Development and Ethnic Conflict in Sri Lanka.* Cambridge: Cambridge University Press.

Verdery, Katherine. 1993. "Whither 'Nation' and 'Nationalism'?" *Daedalus* 122 (3): 37–46.

Verdery, Katherine. 1999. *The Political Lives of Dead Bodies: Reburial and Postsocialist Change*. New York: Columbia University Press.

Vigh, Henrik. 2008. "Crisis and Chronicity: Anthropological Perspectives on Continuous Conflict and Decline." *Ethnos* 73 (1): 5–24.

Visvanathan, Shiv. 1997. *A Carnival for Science: Essays on Science, Technology and Development*. Delhi: Oxford University Press.

Visweswaran, Kamala. 2013. *Everyday Occupations: Experiencing Militarism in South Asia and the Middle East*. Philadelphia: University of Pennsylvania Press.

Vitharana, Vini. *The Night of Doom*. Dehiwela, Sri Lanka: Sarasavi, 1990.

Walker, Rebecca. 2013. *Enduring Violence: Everyday Life and Conflict in Eastern Sri Lanka*. Manchester: Manchester University Press.

Warner, Cameron David, Heather Hindman, and Amanda Snellinger, eds. 2015. "Aftershocked: Reflections on the 2015 Earthquakes in Nepal." Hot Spots, *Fieldsights*, October 14, 2015. https://culanth.org/fieldsights/series/aftershocked-reflections-on-the-2015-earthquakes-in-nepal.

Wax, Emily. 2009. "Sri Lanka's President Rajapaksa Sought to Silence Astrologer Bandara." *Washington Post*, November 17, 2009.

Weerakoon, Gagani. 2020. "11 Years Later: Back on the Frontlines, Fighting for You." *Ceylon Today*, May 23, 2020. https://ceylontoday.lk/news/11-years-later-back-on-the-frontlines-fighting-for-you [link no longer active].

Weisenfeld, Gennifer. 2012. *Imaging Disaster: Tokyo and the Visual Culture of Japan's Great Earthquake of 1923*. Berkeley: University of California Press.

Weisenfeld, Gennifer. 2015. "Imaging Disaster: Tokyo and the Visual Culture of Japan's Great Earthquake of 1923." *Asia-Pacific Journal: Japan Focus* 13 (6): 1–15.

Weliamuna, J. C. 2011. "Lifting of Emergency: Exposing the Sham Exercise." *Groundviews*, September 16, 2011. https://groundviews.org/2011/09/16/lifting-of-emergency-exposing-the-sham-exercise/.

Welikala, Asanga. 2008. *A State of Permanent Crisis: Constitutional Government, Fundamental Rights, and States of Emergency in Sri Lanka*. Colombo: Centre for Policy Alternatives.

Whitaker, Mark. 1997. "Tigers and Temples: The Politics of Nationalist and Non-modern Violence in Sri Lanka." *South Asia: Journal of South Asian Studies* 20 (1): 201–14.

Whitaker, Mark. 1999. *Amiable Incoherence: Manipulating Histories and Modernities in a Batticaloa Tamil Hindu Temple*. Amsterdam: VU University Press.

Whitaker, Mark. 2004. "Tamilnet.com: Some Reflections on Popular Anthropology, Nationalism, and the Internet." *Anthropological Quarterly* 77 (3): 469–98.

Whitaker, Mark. 2007. *Learning Politics from Sivaram: The Life and Death of a Revolutionary Tamil Journalist in Sri Lanka*. London: Pluto.

Wickramasinghe, Kamanthi. 2020. "Sri Lanka's Post-war Defence Budget: A Systematic Review." *Daily Mirror*, February 5, 2020. https://www.dailymirror.lk/print/business-news/Sri-Lankas-post-war-defence-budget-A-systematic-review/273-182511.

Wickramasinghe, Nira. 2009. "After the War: A New Patriotism in Sri Lanka?" *Journal of Asian Studies* 68 (4): 1045–54.

Wickramasinghe, Nira. 2014. *Sri Lanka in the Modern Age: A History*. Oxford: Oxford University Press.

Wickremasekare, Damith. 2011. "Action against Tsunami Scaremongers." *Sunday Times*, April 17, 2011. https://www.sundaytimes.lk/110417/News/nws_04.html.

Wickremasinghe, Suriye. 2010. "Precursors of the PTA." In *Prevention of Terrorism Act (PTA): A Critical Analysis*, edited by Bertram Bastiampillai, Rohan Edirisinghe, and N. Kandasamy. Colombo: Centre for Human Rights and Development.

Wigneswaran, C. V. 2010. "PTA, National Security and Human Rights—a Viewpoint." In *Prevention of Terrorism Act (PTA): A Critical Analysis*, edited by Bertram Bastiampillai, Rohan Edirisinghe, and N. Kandasamy. Colombo: Centre for Human Rights and Development.

Williams, Bianca C. 2017. "#MeToo: A Crescendo in the Discourse about Sexual Harassment, Fieldwork, and the Academy (Part 2)." *Savage Minds*, October 28, 2017. https://savageminds.org/2017/10/28/metoo-a-crescendo-in-the-discourse-about-sexual-harassment-fieldwork-and-the-academy-part-2/.

Williams, Raymond. 1977. *Marxism and Literature*. Oxford: Oxford University Press.

Wilson, A. J. 2000. *Sri Lankan Tamil Nationalism*. Vancouver: University of British Columbia Press.

Wilson, Matthew. 2017. *New Lines: Critical GIS and the Trouble of the Map*. Minneapolis: University of Minnesota Press.

Wimmer, Andreas, and Nina Glick Schiller. 2002. "Methodological Nationalism and Beyond: Nation-State Building, Migration and the Social Sciences." *Global Networks: A Journal of Transnational Affairs* 2 (4): 301–34.

Winslow, Deborah, and Michael D. Woost, eds. 2004. *Economy, Culture, and Civil War in Sri Lanka*. Bloomington: Indiana University Press.

Witjes, Nina, and Philipp Olbrich. 2017. "A Fragile Transparency: Satellite Imagery Analysis, Non-state Actors, and Visual Representations of Security." *Science and Public Policy* 44 (4): 524–34.

Wolfe, Patrick. 2006. "Settler Colonialism and the Elimination of the Native." *Journal of Genocide Research* 8 (4): 387–409.

Wood, Denis, and John Fels. 1992. *The Power of Maps*. New York: Guilford.

Woods, Eric Taylor, Robert Schertzer, Liah Greenfeld, Chris Hughes, and Cynthia Miller-Idriss. 2020. "COVID-19, Nationalism, and the Politics of Crisis: A Scholarly Exchange." *Nations and Nationalism* 26 (4): 807–25.

Wool, Zoë H. 2017. "In-durable Sociality: Precarious Life in Common and the Temporal Boundaries of the Social." *Social Text* 35 (1): 79–99.

Wright, Kelly-Ann, Ilan Kelman, and Rachel Dodds. 2021. "Tourism Development from Disaster Capitalism." *Annals of Tourism Research* 89 (July). https://doi.org/10.1016/j.annals.2020.103070.

Yoneyama, Lisa. 1999. *Hiroshima Traces: Time, Space, and the Dialectics of Memory*. Berkeley: University of California Press.

Yusoff, Mohammad Agus, Athambawa Sarjoon, and Zawiyah Mohd Zain. 2018. "Resettlement of Northern Muslims: A Challenge for Sustainable Post-war Development and Reconciliation in Sri Lanka." *Social Sciences* 7 (7): 1–20.

Zaloom, Caitlin. 2004. "The Productive Life of Risk." *Cultural Anthropology* 19 (3): 365–91.

Zee, Jerry C. 2017. "Holding Patterns: Sand and Political Time at China's Desert Shores." *Cultural Anthropology* 32 (2): 215–41.

Zia, Ather. 2019. *Resisting Disappearance: Military Occupation and Women's Activism in Kashmir*. Seattle: University of Washington Press.

Index

Note: Page numbers in italics refer to figures.

abductions, 71, 74, 154n11
Accelerated Mahaweli Development Project (AMDP), 47
alien others, 41, 48–52, 56, 153n12. *See also* dehumanization
Allah, 3
Amarasuriya, Harini, 139
Amaraweera, Mahinda, 72
American Association for the Advancement of Science (AAAS), 117, 118–21
Amman, Karuna, 12, 16, 48, 65
Amnesty International, 117
Anderson, Benedict, 28
Anthropocene, 7, 61
anthropology, 7, 61, 79–80, 120, 150n21, 151n25
anti-Christian violence, 142
anticipation, 31–32, 59–75, 93, 146
anticipatory states, 32, 64, 69, 74–75
anti-Muslim violence, 142–43, 153n10
anti-Tamil violence, 10, 46, 50
Appu, Puran, 38–39
Aragalaya, xiii–xv, 147n4
arrests, 68, 71–72, 130, 149, 154n11
Arugam Bay, 19

Arulingam, Swasthika, 147n4
astrogeophysics, 73
astrology, 32, 69, 71–73
attrition warfare, 117
Australia, 70
authoritarianism, xiv, 30, 139, 142

Bandara, Chandrasiri, 71
Ban Ki-moon, 121, 124
Bengkulu earthquakes (2007), 65–66
Benjamin, Walter, 28–29, 97
Berlant, Lauren, 25–26, 81–82
biosecurity, 63, 153n4
Black July (1983), 9
Bodu Bala Sena (Army of Buddhist Power or Buddhist Power Force, BBS), 142–43
Bolshevik-Leninists, 148n9
border zones, 23, 27, 63
Boxing Day Tsunami (2004), 4
British Broadcasting Corporation (BBC), 117, 121
British colonialism, 8–9, 39, 49, 103, 105–6
British Crown Colonies, 49
British East India Company, 156n1

Buddhism, 8, 49; in constitution, 9, 48–52; response to colonialism, 9. *See also* Accelerated Mahaweli Development Project (AMDP); *Mahavamsa*; Sinhala Buddhist nationalism
buffer zones, 4, 19, 27, 63, 80, 82–83, 89
"build back better," 27, 83, 155n4
Burma Plate, 4
Buthpitiya, Vindhya, 158n10

caste, 10, 43, 45, 106
catastrophe, 7, 13, 26–27, 29–30, 32, 42, 55, 64, 69, 72, 80; and progress, 97–98
catastrophization, 147n4
Ceasefire Agreement (2002), 12, 16, 21, 23, 48, 115, 149n12
Centre for Policy Alternatives, 51
Ceylon, 9, 49–50, 104–5, 148, 148n9, 156, 156n1
Ceylon Citizenship Act No. 18 (1948), 50
Chandrasena, C. L., 70–71
Channel 4 News (UK), 124–25
checkpoints, 12, 23, 27, 32, 38, 48, 63, 67, 80, 87, 89–90, 100, 139, 141, 154n5
China, 73, 139
Chinese Metallurgical Group Corporation, 139
Chitra Velayudha Swami Temple, 59
Christchurch earthquake (2011), 73
Christianity, 8–9, 142, 153n10; humanitarian organizations, 16
chrononormativity, 29
chronopolitics, 31–32, 80–81, 91
Civilian Safe Zone (CSZ), 94, 116–17, 119, 122
civil war, phases of, 10
civil war as disaster, 21, 139
class, xiii, 10, 43, 47, 50, 52, 148n9, 148n11
climate security, 8
Clinton, Bill, 155n4
clock towers, 98, 99, 156n12
Coastal Conservation Department (CCD), 83
coastal no-build buffer zones, 4
Cold War, 27, 157n6
Colebrooke-Cameron Commission Reforms (1832–33), 105–6

collective memory, 130
Collier, Stephen J., 153n2
Colombo, xiv, 16, 21, 37, 55, 72, 74, 84, 90, 111; disappearances in, 88; disaster management in, 23, 60–61, 65; postwar life in, 139; protests in, 124; violence in, 46, 143; war's end in, 25–26; Year for War billboards in, 16–17, *17*
colonialism/imperialism, 7, 53, 107, 113; British, 8–9, 39, 49, 103, 105–6; Dutch, 39, 103–6, 156n1; Israeli, 55, 156n7; Portuguese, 8, 39, 103–6; state-sponsored, 20, 31, 43, 45–47
Communists, 148n9
counterinsurgency, 7, 55
counterterrorism projects, 8, 9, 12, 31, 50–53, 55, 146
COVID-19 pandemic, xiii, 143–46, 150n22
Crampton, Jeremy W., 158n11
Criminal Investigation Department (CID), 72
crisis, 13, 99, 137–38, 145, 147n3; and disaster, 7, 83; as ordinary, 30, 80, 82; permanent, 42; and temporality, 28
critical geography, 113, 157n6
cyclone (1978), 31, 41–43, 46–48, 53, 89, 152n2

Daily Mirror, 70, 73, 74
d'Angelo, Jacopo, *104*
Datiya, 39
Dawdy, Shannon, 151n32
Debord, Guy, 103, 120
Deepan, 127
Defense Production Act (US, 1950), 146
dehumanization, 52–53, 133. *See also* alien others
Deleuze, Gilles, 79, 81, 153n2, 156n9
democracy, xiv, 27, 49, 144, 148n9
dengue, 55
De Siva, Marisa, 147n4
development, 45, 47, 63, 116, 139, 153n11, 157n4; disaster risk reduction in, 18; post-tsunami, 15, 20; postwar, xiv, 20, 25; for tourism, 19. *See also* Reconstruction and Development Agency (RADA); Sri Lankan Urban Develop-

ment Authority; United Nations Development Program (UNDP)
Dharmaratnam, Sivaram "Taraki," 154
Dhatusena, King, 39
Dillard, Annie, 1
disappearances, 71, 74, 87, 89, 154n11. *See also* white van culture
disaster as analytic, 4–7
disaster capitalism, 5, 19–20
disaster governance, 6, 18, 61, 79
disaster management, 4, 6, 9, 12–17, 21, 23–25, 40, 52–55, 59–61, 69–74, 83, 100, 109, 113, 146, 153n2. *See also* Disaster Management Act No. 13 (2005); Disaster Management Centre (DMC); disaster mitigation; disaster risk management (DRM); disaster risk reduction (DRR)
Disaster Management Act No. 13 (2005), 8, 13, 15, 55, 144, 149n13
Disaster Management Centre (DMC), 13, 25, 60–61, 65–66, 71, 73–74, 113, 127, 140–41
disaster mitigation, 13, 20, 33. *See also* disaster management
disaster nationalism, definition, 4–5
disaster risk management (DRM), 6, 13, 15, 28, 52, 61, 63–64, 113, 146, 149nn13–15
disaster risk reduction (DRR), 6, 13, 15, 18, 33, 52, 53, 61, 113, 146, 149n13, 149n15
Disaster Risk Reduction and Climate Variability symposium (2009), 18, 20, 53
disaster warning systems, 4, 13, 25, 27, 60–61, 63, 73, 92, 127
disaster warning towers, 29, 60–61, 62, 65–67, 75, 90, 127, 141, 153n1
displaced populations, 4, 12, 18–19, 23–24, 48, 64, 82, 109, 114, 140; of Muslims, 152n3
dissent, 71–72, 131, 154n11
Downey, Gary Lee, 37, 41
Dry Zone, 43–45, 153n11
Dumit, Joseph, 37, 41
Dutch Burghers, 8
Dutch Ceylon, 105
Dutch colonialism, 39, 103–6, 156n1

Dutch East India Company (VOC), 104
Dutugemunu, King, 38–39

Early Warning Capacity Building, 127
earthquakes, 15, 19, 23, 27, 61, 65–66, 70–73, 75, 107, 149n15, 154n8, 155n12
Easter bombings (2018), 143
Eastern Province, 8, 24, 64–65, 83, 86, 94, 145; Ampara (town), 46, 88, 113, 127; Ampara District, 19, 43, 43–45, 44, 111; Batticaloa District, 43, 44, 46, 57, 80, 111, 140, 152n2; Kalmunai, 25, 43, 45, 64, 79, 84, 87, 101, 127, 130–31, 133, 138; Maruthamunai, 95; Thirrukkovil, 59
eclipses, 69–70
economic liberalization, xiv, 47, 99
Eelam (imagined Tamil homeland), 10, 17, 128
Eelam III (1994–2001), 10
Eelam IV (2005–2009), 10, 12
Eelam War I (1983–1987), 10
Eelam War II (1987–1993), 10
Eknaligoda, Prageeth, 154n11
Elara, 39
emergence, 29, 31, 37–56, 69, 100, 106, 108, 128, 132, 138
endurance, 6–7, 21, 29, 32–33, 69, 73–75, 79–100, 137, 140, 146; term, 155n1
English language, 8, 50, 61, 70, 73, 106
ethnicity, xiii, 8, 20, 49, 148nn8–9, 153n3; and civil war, 9–10, 47, 153n3; in Colombo, 139; and COVID-19 pandemic, 146; and LTTE, 10, 24, 44–47, 154n6; and Sinhalese Buddhist nationalism, 39, 49, 52; and tsunami aftermath, 43, 83
ethnography, 21–25, 29, 32, 80, 107
evacuation, 73–74, 101, 140; drills, 13, 25, 59, 61, 63, 65, 69, 71
event simulations, 13, 63
extrajudicial killings, 12

Fajiana, 95–96, 132–33
fantasy. *See* national fantasy
Farood, 18–19, 89, 127, 131
Fathima, 66

INDEX **193**

Fawzi, 68
Fazil, M. M., 45
field notes, 32, 80
fieldwork, 4, 16, 21–25, 31, 88, 113, 139; and racialization, 150n21
Fonseka, Sarath, 154n11
"forget the past," 99, 133
Fortun, Kim, 147n3
Foucault, Michel, 151n32, 153n2
France, 105, 121
Free Aceh Movement/GAM (Gerakan Aceh Merdeka), 150n16

G11 (Group of Eleven) meetings, 92, 116
Galle Face Green, xiv, 37
Gal Oya Irrigation Scheme (1949), 45
Gal Oya River, 45–46
Gandhi, Rajiv, 149n12
Ganeshananthan, V. V., 35, 57, 77, 101
geo-bodies, 106–8, 131–32, 156n2
Geographic Information Systems (GIS), 107–8, 113, 117, 157n3
global governance, 7
Global Platform for Disaster Risk Reduction, 18
Global War on Terror, 40
Gnanasara, Galagoda Aththe, 142
Govindrajan, Radhika, 153n10
grease devil, 137–38
Great Britain, 109, 121, 124; British colonialism, 8–9, 39, 49, 103, 105–6. *See also* United Kingdom
Great Kanto earthquake (1923), 107
Greenhouse, Carol J., 151n25
ground truth, 33, 117, 132, 154n6
Groundviews.org, 72
Guantanamo Bay Naval Base, 53
Guterres, António, 145

Haiti earthquake (2010), 7
Hall, Stuart, 42, 138
Haniffa, Farzana, 147n4
Hartman, Saidiya, 33
Hasbullah, Shahul, 54
Hawai'i, 61
Hewitt, Kenneth, 6
High Security Zones, 48

Hinduism, 8, 60; Hindu nationalism, 146, 153n10
Holmes, John, 121
homogeneous time, 28–29, 79, 99
Hoole, Rajan, 148n9
housing schemes, 4, 32, 82–89, 141
humanitarianism, 15, 25, 92–93, 107, 109, 113–14, 116–17, 119–22, 133; Christian, 16; digital, 158n7; golden wave, 19; militarized, 7
"human-made" disasters, 15, 18, 27, 53, 100, 147n4
human rights, 47, 51–53, 116–17, 121–22, 124–25, 145. *See also* humanitarianism
Human Rights Watch, 117
Hungary, 146
Hurricane Harvey, 148n6
Hurricane Ida (2021), 7
Hurricane Katrina (2005), 7, 151n32
Hurricane Maria (2017), 7
Hyogo Declaration (2005), 13
Hyogo Framework for Action (2005–2015), 13–15

immunology, 55–56
Independence Day, 17
India, 5, 49–50, 70, 73, 146, 153n10; South, 8, 20, 39, 49, 155n4
Indian and Pakistani Residents' (Citizenship) Act No. 3 (1949), 50
Indian Army, 47
Indian/Malaiyaha/Hill Country Tamils, 8, 49–50, 148n9
Indian Ocean, 4–5, 44, 70, 73, 103
Indian Ocean earthquake (2004), 3–5, 5, 7, 24, 65, 155n4
Indian Peacekeeping Force (IPKF), 12, 47, 149n12
India Plate, 4
Indonesia, 3, 5, 66, 70, 72, 130, 150n16, 153n4; Aceh, 4, 65, 73, 150n16; Ach, 4
Indo–Sri Lankan Accord, 149n12
infrastructure, xiv, 27, 32, 42, 63, 80, 99, 153n4, 154n5; Chinese-built, 139; disaster, 7, 13; tourist, 19; war-related, 63

infrastructures of feeling, 154n10
insecurity, 6, 21, 28, 30–32, 40, 61, 69, 75, 79–81, 83, 100, 107, 154n5
Internally Displaced Persons (IDPs), 12, 109, 117. *See also* displaced populations
international aid, 4, 15–16, 19, 25, 72, 109, 113–14, 117, 150n16, 155n4
International Center for Emergency Techniques, 127
International Commission of Jurists, 52
International Committee of the Red Cross, 85, 93, 117
International Water Management Institute (IWMI), 109
internment camps, 24, 65
Iranathivu, 145
Ismail, S. M. M., 45
Israel, 146; Israeli colonialism, 55, 156n7
Israeli Security Agency, 146
Italy, L'Aquila, 18, 155n12
Ives, Sarah, 53

Jaffna Youth Congress, 49
Jameson, Fredric, 28
Janatha Vimukthi Peramuna (JVP, People's Liberation Front), 47, 74; and Indo–Sri Lankan Accord, 149n12; 1971 insurrection, 12, 51, 154n11; 1987 insurrection, 12, 47, 148n11, 149n12, 154n11
Japan, 70, 107; Sendai, 73; Triple Disaster (2011), 7, 73
Japanese Meteorological Association, 61
Jathika Hela Urumaya (JHU, National Heritage Party), 158n8
Jayasuriya, Jagath, 55
Jayawardena, J. R., 41, 46, 149n12, 151n1
Jayawardena, Kumari, 148n9
Jegathesan, Mythri, 50
Jordan, 92, 94
journalism, 33, 71, 92, 117, 119, 120–21, 154n11

Kadirgamar, Ahilan, 10, 147n3
Kandyan kingdom, 103
Kaplan, Caren, 157n6
Karaitivu army base, 88
Kataragama, 59
Kelly, John, 29
Kelman, Ilan, 150n20
Keppetipola, 38–39
Khalili, Laleh, 55
Klein, Naomi, 5, 19–20
Klima, Alan, 125
Knowles, Scott, 28, 148n6
Korf, Benedikt, 54
Kouchner, Bernard, 121
Krygier, John, 158n11
Kumaratunga, Chandrika Bandaranaike, 12, 15–16, 82, 114

Lakoff, Andrew, 153n2
Lakshmi, 66–67, 97, 130
Lal, Vinay, 51
L'Aquila earthquake (2009), 18–19, 155n12
lateral agency, 100
Latour, Bruno, 54
Lessons Learnt and Reconciliation Commission, 124
Liberation Tigers of Tamil Eelam (LTTE), 26, 46, 50, 132, 140, 148n10; expulsion of Muslims, 152n3; imagined mono-ethnic homeland, 8, 11, 24, 44–45, 65, 87, 154n6; Karuna Amman's breakaway from, 12, 16, 65; Prabhakaran's death, 68, 127–31; and terrorism, 40, 92; and tsunami aid, 15–16, 114, 150n16; war with government, 4–5, 8, 10–12, 17, 20–24, 26, 37–38, 40, 47–48, 55, 80, 87, 91–96, 115–17, 119–20, 124, 131, 133, 141, 149n12
Loeb, Zachary, 148n6
Lowe, Celia, 153n4
Lutz, Catherine, 147n1

Mahavamsa, 106
Malaysia, 70
Maldives, 70
Maldonado, Julie K., 20
"man-made" disasters. *See* "human-made" disasters
MapAction, 109, 110

INDEX 195

maps/mapping, 17, 29, 33, 44, 70, 103–21, 126, 128, 132, 156n2, 157n6, 157nn3–4, 158n11
Martin, Emily, 55
Marxism, 51, 74, 148n11
Masco, Joseph, 27, 54
Mauritius, 70
Mayadunne, King, 39
McGilvray, Dennis, 45, 152n2
metasigns, 108, 115
methodological nationalism, 151n25
methodology of book, 4–7, 21–25. *See also* disaster as analytic; ethnography; field notes; fieldwork
Michael, George, 133
Miliband, David, 121
militarized humanitarianism, 7
Mohamed, 66, 141
Mohan, 89–91, 100
Mufeetha, 3
Muslim communities, 6, 8, 68, 81, 92, 97, 106, 131–32, 137–39, 145–46, 150n22; anti-Muslim violence, 142–43, 153n10; displacement of, 152n3; in the east, 23, 26, 45, 83, 87; in Kalmunai, 25, 42–43; and LTTE, 48, 154n6; in Maruthamunai, 95; and Prevention of Terrorism Act, 51; and P-TOMS, 15; and Sinhala Buddhist nationalism, 9, 141–43

Nanthikadal Lagoon, 128
National Building Resource Organization, 141
national fantasy, 21, 25–28
National Movement Against Terrorism, 158n8
National Operation Centre for Prevention of COVID 19 Outbreak (NOCPC), 144
National Safety Day, 140–41
national securitization, 6, 21, 40, 53
national security, xiv, 13, 29, 40, 49, 51–53, 80, 100, 126, 139, 141, 143, 146
National Tribute to Security Forces, 37
National Victory Day Parade (2009), 35
Nations and Nationalisms, 146

neoliberalism, 19–20, 47
Nepal earthquake (2015), 7
Nervous System, 64, 69
Netherlands: Dutch colonialism, 39, 103–6, 156n1; French invasion of, 105
New Zealand, Christchurch, 73
Nitharshini, 24
No Fire Zone (NFZ), 93–94, 117
nongovernmental organizations (NGOs), 21, 150n16
nonhuman, 53
nonviolence, xiv, 10
Northern Province, 8, 24; Jaffna, 10, 45–46, 48–49, 95, 111, 133, 138, 152n3, 158n10
Norway, 12

objectivity, 26, 107–8, 113–14, 119, 157n6
Occupied Palestinian Territories, 55, 114, 147n4, 156n7
Ophir, Adi, 147n4
ordinariness, 29–30, 80–82
Orr, Jackie, 151n24

Pacific Basin, 73
Pacific Ocean, 69
Pacific Tsunami Warning Center, 61
Pada Yatra, 59
Palayamara, 39
Palestine, 55, 114, 147n4, 156n7; counterinsurgency in, 5
pandemic nationalism, xiii, 143–46
pandemics, xiii, 8, 13, 63, 143–46. *See also* COVID-19 pandemic
Parliamentary Elections Amendment Act (1949), 50
Parvati, 68
Peebles, Patrick, 152n6
Peters, Laura E. R., 150n20
Philippines, 70
Pitiya, 39
Polo, Marco, 106
Portugal, 8; Portuguese colonialism, 8, 39, 103–6
postcolonialism, 9, 106, 138, 152n7, 153n11
postindependence politics, 31, 41, 45, 50, 52

post-tsunami development, 20
Post-Tsunami Operation Management Structure (P-TOMS), 15–16, 114–15, 150n17
Povinelli, Elizabeth, 29, 80
Prabhakaran, Velupillai, 12, 39, 68, 127–31, 133, 158n9
precarity, 6, 21, 30, 54, 69, 146, 148n11
Premadasa, Ranasinghe, 99, 149n12
Premalal de Silva, Ranjith, 117
Presidential Task Force for Archaeological Heritage Management in the Eastern Province, 145
Presidential Task Force to Build a Secure Country, Disciplined, Virtuous and Lawful Society, 144–45
Prevention of Terrorism Act (PTA), 41, 47, 51–52, 142, 154n9
Prevention of Terrorist (Proscription of Extremist Organizations) Regulations No. 2 (2021), 52
promiscuous commingling, 107
Proscribing of the Liberation Tigers of Tamil Eelam and Similar Organizations Act No. 16 (1978), 41, 50–51
Ptolemy, 103, *104*, 106
Public Security Ordinances, 154n9
purity/purification, national, 19, 21, 40, 49, 54–56, 67
Puttalam, *111*, 152n3

Qatar, 68

race, xiii, 7, 20, 50, 81, 153n12; and colonialism, 9, 53
racism, 10
Raheem, 43, 66
Rajakarunayake, Lucien, 71
Rajapaksa, Basil, 144
Rajapaksa, Gotabaya "The Terminator," xiii–xiv, 52, 139, 143–45, 147n2, 154n11
Rajapaksa, Mahinda, xiv, 16, 18, 23, 25–27, 33, 37–40, 52, 54, 56, 71–72, 83, 91–92, 96–97, 99, 116, 139–40, 142–43

Rajasingha I, King, 39
Rajchman, John, 156n9
Ramaswamy, Sumathi, 158n9
Rameez, M. A. M., 45
Rani, 84
Ravi, 68, 89
Reagan, Ronald, 151n1
Reconstruction and Development Agency (RADA), 16
Red Cross. *See* International Committee of the Red Cross
Reddy, Elizabeth, 154n8
rehabilitation, 13, 33
reiteration, 18–19, 32–33, 103–33
Reporters without Borders, 71
resettlement, 18, 23, 43, 53, 64, 83, 109, 152n3, 154n6; of displaced Muslims, 152n3
resilience, 155n1
restructuring, 4, 19–20, 47, 63
Rheinberger, Hans-Jörg, 126
Rubble Road, *77*
Rupavahini, 128
Ruwanpura, Kanchana, 156n6
Ruzfina, 96

Safety Sri Lanka exhibit, 140–41
Salowdeen, 94
Samarasinghe, Mahinda, 18–20, 53, 64
Sanjana, 84
satellite imagery, 60, 107, 117, 119–22, 132, 145, 157n6. *See also* Geographic Information Systems (GIS)
Schiller, Nina Glick, 151n25
Schonthal, Benjamin, 20
Schuller, Mark, 20
science and technology studies, 61
Scott, David, 106
Scott, James, 157n4
securitization, xiv, 6, 12, 21, 26–29, 31, 40, 52, 53–54, 63–64, 87, 141, 144, 146
Security Force Headquarters of the East, 140
security-scapes, 8
Selvi, 42–43, 48, 85–86, 89, 140
Senanayake, D. S., 45, 148n9
Senaratne, Atula, 72–73

Senayayake, D. S., 49
Sendai earthquake (2011), 73
Sendai Framework for Disaster Risk Reduction (2015–2030), 6
September 11, 2001, attacks, 28
sexual violence, 90, 95, 149n12
Sharma, Sarah, 29
Sharpe, Christina, 138
Short Message Service (SMS), 60, 116
Silva, Shavendra, 144
Silva, Kalinga Tudor, 153n11
Sinbad, 106
Singapore, xiv, 70
Sinhala Buddhist nationalism, 15, 20–21, 45, 47–49, 52, 54, 145, 148n11, 149n12, 154n6, 158n8; and civil war, 8–9, 133, 141–43; disaster management upholding, xiv–xv, 6, 31, 139–40; J. R. Jayawardena on, 151n1; Mahinda Rajapaksa on, 39–41; and tsunami aid, 15–16
Sinhala Kingdom, 38
Sinhala language, xiv, 8, 61
Sinhala Only Act (1956), 46, 50
Sinhalese communities, 8–10, 45–46, 49, 71, 106, 127, 148nn8–9, 153n11; and P-TOMS, 15–16; in the south, 25–26, 83. *See also* Sinhala Buddhist nationalism
Sinhalese Lion, 26
Sinhalese New Year, 26
Sinhalization, 10, 47, 141
Sirisena, Maithripala, 52, 142–43
Sitamma, 68, 86–89, 97, 130
Siva, 39
Sivaram, Taraki, 154n11
slow death, 81–82
slow life, 32, 80–82, 91
Snopes.com, 70
Sontag, Susan, 130
Southern Province, 8; Hambantota, 111, 140
Special Task Force (STF), 25, 43, 48, 67–68, 87, 90, 94–95, 97
spectacle (Debord), 103, 120
speculative life sciences, 63
Spencer, Jonathan, 47, 54, 139

Sri Lankan Air Force, 35, 37
Sri Lankan Armed Forces, 141, 143–44
Sri Lankan Army, 12, 20, 37, 43, 55, 67, 80, 88, 90, 95, 97, 131, 140–41, 143, 144–45; attacks on civilians, 93–95, 116–17, 133; disappearing people, 89, 154n11
Sri Lankan Constitution, 144; Eighteenth Amendment, 142; and nationalism, 48–52; 1972 version, 9; 1978 version, 31, 41, 46; Thirteenth Amendment, 149n12
Sri Lankan Department of Census and Statistics, 148n7
Sri Lankan Geological and Mines Bureau, 71
Sri Lankan Government News Portal, 140
Sri Lankan Independence, xiii, 8–9, 16–17, 49, 148n9. *See also* postindependence politics
Sri Lankan Information Department, 92
Sri Lankan Medical Association, 145
Sri Lankan Meteorology Department, 66, 70, 71, 73, 141
Sri Lankan Ministry of Defence, 32, 46, 80, 92, 93, 139, 141, 143; *Lies Agreed Upon*, 125, 126; "When the Camera Lies for Terror," 117, 119, 122
Sri Lankan Ministry of Telecommunication and Information Technology, 156n10
Sri Lankan Navy, 141
Sri Lankan Parliament, 24, 37, 39, 41, 46, 50, 52, 142, 148n9, 149n13, 152n2; dissolving of, 144–45; Parliamentary Committee on Natural Disasters, 13; "Towards a Safer Sri Lanka," 13, 60
Sri Lankan Supreme Court, 16, 114
Sri Lankan University of Peradeniya, 72
Sri Lankan Urban Development Authority, 139, 141
Sri Lanka's Killing Fields, 124, 125
Srinivas, Smriti, 155n3
Sri Wickrama Rajasinghe, 38
state colonization, 20, 31, 43, 45–47
statecraft, 4, 21, 139, 141

198 INDEX

state of emergency, xiv, 9, 12, 24–25, 40, 42, 47, 51, 67, 87, 154n7, 154n9
Stewart, Kathleen, 29–30
Strassler, Karen, 108, 114, 130
Subramaniam, Banu, 153n12
suicide bombings, 12
Sunda Trench, 3, 65
surveillance, 87, 90, 108, 120
Switzerland, Geneva, 18, 53, 64

Tamil communities, 6, 42–43, 68, 81, 83, 127, 137–38, 145, 150n22, 152n2; Ceylon Tamils, 49–50; and civil war, 8–12, 18, 38–39, 48, 93, 141–42; internment of, 18, 24, 65, 154n7; Indian/Malaiyaha/Hill Country Tamils, 8, 49–50, 148n9, 152n7; and PTA, 51; and Sinhala Buddhism, 45–46, 106, 133; state disappearances of, 87–90, 154n11; Tamil nationalism, 50, 130, 154n6, 156n10, 158n9; and the tsunami, 23, 92. *See also* Liberation Tigers of Tamil Eelam (LTTE)
Tamil Eelam, 8, 10, 11, 16, 87, 130, 132. *See also* Eelam (imagined Tamil homeland)
Tamil language, 8, 24, 26, 38, 61, 87; legal restrictions on, 50; and *tamilparru*, 158n9
Tamil New Tigers, 10
Tamil United Liberation Front (TULF), 6, 10, 46
Taprobane, 103, *104*
Task Force to Rebuild the Nation (TAFREN), 16
temporality, 9, 39–40, 80, 151n28, 155n3; and disaster, 6–7, 13, 28–31, 33, 61, 79, 86, 90, 100, 107; homogeneous time, 28–29, 79, 99; preemptive, 28–29, 53, 61, 100, 146, 153n2. *See also* anticipation; chronopolitics; emergence; slow life; waiting
Thailand, 5, 20
Tharanga, 89
Thiranagama, Sharika, 128, 130, 137, 143–44, 150n22
Thongchai, Winichakul, 108, 156n2

Time magazine, 91
Tissainayagam, J. S., 154n11
tourism, 19, 114, 133
transitional justice, 142
Trincomalee District, 43, 44, 66, *111*
Trotskyists, 148n9
tsunami lung, 155n5
Turkey, 73
200 Garment Factory Program (200 GFP), 99

unassimilated others, 41, 50
undomiciled others, 41
United Kingdom, 51. *See also* Great Britain
United Nations (UN), 117, 120–21, 124, 145, 155n4
United Nations Development Program (UNDP), 15, 61, 65, 127
United Nations General Assembly, 15
United Nations Humanitarian Information Centre (UNHIC), 109
United Nations Human Rights Council (UNHRC), 142
United Nations International Decade for Natural Disaster Reduction, 149n15
United Nations International Strategy for Disaster Reduction (UNISDR), 15, 149n15
United Nations Office for Disaster Risk Reduction, 6
United Nations Office for the Coordination of Humanitarian Affairs (UNOCHA), 109, 113, 140
United Nations World Conference on Disaster Reduction (2005), 15
University Teachers for Human Rights (Jaffna, UTHR-J), 48
US Civil War, 27
US Department of Defense, 157n6
Usha, 24, 86

Valagamba, King, 39
Valluvan, Sivamohan, 48–49
Venugopal, Rajesh, 47, 137
Verdery, Katherine, 126–27, 152n5

Vereenigde Nederlandsche Geoctroyeerde Oostindische Compagnie (Dutch East India Company, VOC), 104
Vijayabahu, King, 39
Vimaladharmasuriya, King, 39
vital systems security, 63, 153n2
Vitharana, Vini, 152n2
vulnerability, 7, 13, 15, 19, 27, 30, 54, 67, 82, 144, 146, 149n15

waiting, xiv, 29, 32, 38, 59–60, 69, 80, 82, 84, 89–90, 95, 99–100
Walker, Becky, 80
weather, the (Sharpe), 138
Weisenfeld, Gennifer, 107

WhatsApp, xv
Whitaker, Mark, 154n11
whiteness, 25, 150n21
white van culture, 71–72, 138, 154n11. *See also* disappearances
Wickrematunga, Lasantha, 154n11
Wickremesinghe, Ranil, xiv, 143
Wimmer, Andreas, 151n25
World Health Organization (WHO), 145
World Trade Center, 28
World War II, 81

xenophobia, 49, 153n12

Year for War (2008), 16–17, 17, 48, 115